T4-AHF-314

PRENTICE-HALL

Foundations of Economic Geography Series

NORTON GINSBURG, *Editor*

GEOGRAPHY OF MANUFACTURING, *Gunnar Alexandersson*

GEOGRAPHY OF MARKET CENTERS AND RETAIL DISTRIBUTION, *Brian J. L. Berry*

GEOGRAPHY OF AGRICULTURE, *Howard F. Gregor*

ENERGY IN THE PERSPECTIVE OF GEOGRAPHY, *Nathaniel B. Guyol*

MANUFACTURING LOCATION, *James Lindberg*

GEOGRAPHY OF URBAN LAND USE, *Harold Mayer*

GEOGRAPHY OF WATER RESOURCES, *W. R. Derrick Sewell*

GEOGRAPHY OF TRANSPORTATION, *Edward J. Taaffe* and *Howard L. Gauthier, Jr.*

GEOGRAPHY OF INTERNATIONAL TRADE, *Richard S. Thoman* and *Edgar C. Conkling*

THE MERCHANT'S WORLD, *James E. Vance, Jr.*

A PROLOGUE TO POPULATION GEOGRAPHY, *Wilbur Zelinsky*

Foundations of Economic Geography Series

UNIVERSITY OF PITTSBURGH LIBRARY
AT BRADFORD

2 95

Geography of Agriculture

THEMES IN RESEARCH

6 ABC
B 94

HOWARD F. GREGOR
Professor of Geography
University of California, Davis

8/26/71
28093

PRENTICE-HALL, INC., Englewood Cliffs, N.J.

© *Copyright 1970 by Prentice-Hall, Inc.*
Englewood Cliffs, N.J.

All rights reserved. No part of this book may be reproduced in any form or by any means without permission in writing from the publisher. Printed in the United States of America.

Library of Congress Catalog Card No.: 73–127855

13–351320–3

Current printing (last number):

10 9 8 7 6 5 4 3 2 1

PRENTICE-HALL INTERNATIONAL, INC., *London*
PRENTICE-HALL OF AUSTRALIA, PTY. LTD., *Sydney*
PRENTICE-HALL OF CANADA, LTD., *Toronto*
PRENTICE-HALL OF INDIA PRIVATE LTD., *New Delhi*
PRENTICE-HALL OF JAPAN, INC., *Tokyo*

Foundations of Economic Geography Series

Among the various fields of geography, economic geography, perhaps more than any other, has experienced remarkable changes within
he past twenty years—so many that it is almost impossible for one scholar
o command all aspects of it. The result has been increasing specializa-
ion on the one hand and, on the other, a fundamental need for bringing
he fruits of that specialization to students of economic geography.

The *Foundations of Economic Geography* Series consists of several
volumes, each focusing on a major problem in economic geography. It is
lesigned to bring the student, whether novice or more experienced, to
he frontiers of knowledge in economic geography, and in so doing, force-
ully to demonstrate the methodological implications of current research
–but at a level comprehensible even to those just becoming aware of the
ascinating problems in the field as it is developing today.

Each volume stands as a contribution to understanding in its own
right, but the series as a whole is intended to provide a broad cross-section
of on-going research in economic geography, stemming from concern
with a variety of problems. On the other hand, the series should not be
regarded as a complete synthesis of work in economic geography, al-
hough the volumes explore in depth certain major issues of keenest
nterest to economic geographers and others in related fields to a degree
impossible in textbooks that attempt to cover the entire field. At the same
time, the student is brought face-to-face with the kinds of intellectual
and conceptual problems that characterize economic geography in a way
that no over-all survey can accomplish. Each volume thus provides a
basis for an intensive exploration of issues that constitute the cutting edge
of research in this most dynamic and demanding field of knowledge.

As time goes on and new volumes appear in the series, the original
volumes will be modified in keeping with new developments and orientations, not only in economic geography, but in the field of geography as a
whole. The first volume to appear in the series, Wilbur Zelinsky's *A Prologue to Population Geography*, acts as a bridge between economic and
cultural geography and as a means for exploring ideas and methods
concerning a problem of increasing interest to geographers and social
scientists alike: the growth, diffusion, and distribution of populations
throughout the world. Brian J. L. Berry's *Geography of Market Centers
and Retail Distribution* attempts to fill a major lacuna in the literature of
economic geography, as it lays down principles concerning the spatial
distribution and organization of marketing in both advanced and lesser
developed economies. In so doing, it provides a bridge between the
geographies of consumption, production, and cities, and links them for

perhaps the first time effectively through a theoretical system, still primitive, but far in advance of comparable formulations.

Gunnar Alexandersson's *Geography of Manufacturing* reflects the need for considering the historical-ecological settings within which manufacturing enterprises originate and flourish. Though superficially nontheoretical, it contains flashing insights into the extreme socioeconomic complexities that have resulted in the world pattern of manufacturing, and it is concerned with an interpretation of that pattern through an evolutionary-descriptive technique applied to selected industries and regions.

Richard S. Thoman and Edgar C. Conkling in their *Geography of International Trade* deal with a topic that has not been given the attention it deserves by economic geographers. By careful analysis of trade data and an imaginative use of graphic and tabular devices, they interpret the pattern and structure of international trade in terms of current monetary and economic blocs. The result is the first modern treatment of one of the basic types of international relations, and thereby an important contribution to the political as well as economic geographical literature.

Unusual for one of a series of this kind, James E. Vance, Jr.'s *The Merchant's World: The Geography of Wholesaling* makes a substantial original contribution to knowledge through the examination of principles applying to the distribution and organization of wholesaling activities in the past several hundred years. Although readily comprehensible to literate neophytes in economic geography, Professor Vance's thesis suggests a major dimension in the geographic study of economic activity, which has been seriously neglected so far. Moreover, he raises fundamental and intellectually demanding questions about the state of theory in economic geography, and his work illustrates the value of an historical–analytical approach to geographic problems.

In his *Geography of Agriculture: Themes in Research,* Howard Gregor has methodically and imaginatively analyzed the intellectual history of this subdiscipline of economic geography. His bibliography alone is a substantial achievement. Even more important, he has presented with extraordinary clarity the major geographical problems relating to the study of agriculture, all in simple, straightforward prose, but at a level of intellectual discourse of as great interest to professional geographers as to novices in the field. He also relates the results of research in agricultural geography to real problems in the nonacademic world.

The other volumes in the series, whether concerned with transportation, urban land uses, manufactural location, or resources management, display "bridging" qualities that transcend the narrow limitations of ordinary descriptive handbooks. All are concerned with the new and the fresh as well as the traditional, and with the transformation of a somewhat parochial field of scholarship into one that is interdisciplinary as well as innovative and pioneering.

NORTON GINSBURG

The University of Chicago

Preface

A book is often introduced with the complaint that too little attention has been paid to its subject. This one cannot be, for more has been written on agricultural geography in one way or another than on any other geographic subfield. What does seem to be in short supply is a survey of the progress in research. Texts on agricultural geography embody much of this research, to be sure, but they are few and all too often they convey the impression of a universally accepted body of information, and not a record of discoveries, ideas, and debates. This last picture is the one I hope to present.

An approach of this kind also demands its price, particularly in this small volume. Only the more important paths of inquiry can be pointed out, and these only briefly described. Documentation can partly remedy this briefness, but it, too, must be severely limited because of the sheer mass and variety of contributions; where there seemed little chance of distorting the overall research picture, citations were further restricted to the North American literature with emphasis on that continent.[1] Handicaps like these have been cheerfully accepted, however, in the hope that both colleagues and students may still benefit: the professionals, from seeing their contributions in a wider perspective; the students, from seeing agricultural geography more as the intellectual adventure that it is.

I have no problem with qualifications in acknowledging the skill of Professor Norton Ginsburg, the editor of this series, in helping make this a much more readable book than one on research trends might usually be. Mr. Robert Rugg, of the Department of Geography, University of Chicago, also deserves full recognition for translating my manuscript maps into products suitable for the printed page. No less heartening has been the generosity of numerous colleagues, geographical societies, and publishing houses in providing me with a wealth of illustrative materials from which to choose the few adapted for presentation here.

HOWARD F. GREGOR

[1] For the additional and particularly abundant illustrations in German and French, one should consult especially the bibliographies appearing in the books of Otremba, Schwarz, Faucher, and George, all noted in this book.

Contents

PART 1 *the consideration of the field*

ONE
Purpose and position *[illegible]*

TWO
Research methods *17*

PART 2 *the study of landscape*

THREE
The role of environment *31*

FOUR
Spatial organization *57*

FIVE
The cultural impress *72*

SIX
Political reflections *84*

SEVEN
The historical context *94*

PART 3 *the search for regions*

EIGHT
Regional types and patterns *111*

NINE
Regional boundaries *130*

PART 4 *the concern for resources*

TEN
Resource destruction *139*

ELEVEN
Population and food supply *151*

Index *165*

PART 1 *the consideration of the field*

CHAPTER ONE

Purpose and position

Old Subject: Young Field

What do agricultural geographers consider to be the objectives of their field, and how do they defend these objectives as unique among those of other fields? Considering that agriculture has dominated the human landscape and claimed most of man's waking hours for several thousand years, specific answers to these questions have been offered surprisingly late. The tardiness reflects the recency of agricultural geography as an organized field of study. Certainly one can find extensive writings by early Greek and Roman writers—Strabo, Caesar, Pliny, and others—on agriculture in Gaul and Germania, but these were neither complete nor systematic. Nor were any significant advances made toward a systematic treatment of agricultural geography in the Middle Ages. Only toward the end of the eighteenth century can one begin to discern a truly geographic literature on agriculture. Then, instead of subscribing to a raw description, researchers began to consider the role of physical conditions in the variegation of agricultural areas. Especially representative of this period are the accounts by the agricultural "topographers," Arthur Young of England and J. N. Schwerz of Germany.[1]

In the nineteenth century, analysis began to give hints of the regional approach which geographers were later to use extensively. The combining of crops and other parts of the areal agricultural complex is reflected by the use of the word "painting" (*Gemälde*) by Alexander von Humboldt in his 1807 reference to the "natural paintings" of tropical lands. Not until almost the end of the century, however, did anything

[1] See, e.g., Young, *The Farmer's Tour Through the East of England*, 4 vols. (London: Strahan, 1770); and Schwerz, *Beschreibung der Landwirtschaft im Niederelsass* (Berlin: Parey, 1816).

appear that resembled a comprehensive agricultural geography, and it was restricted to a statistical study of crop regions in North America.[2] Serious discussions on the nature of agricultural geography were to await the present century, heralded by Krzymowski's article on the "scientific position of agricultural geography" in 1911. By 1933, agricultural-geographic writings had increased to the point of encouraging Leo Waibel to become the first geographer to devote a methodological work to the "problems of agricultural geography."[3]

Purpose and Position

Perhaps it is because such a short period of time has been devoted to methodological discussion that there is almost unanimous agreement on the primary object of study in agricultural geography: the areal variation of agriculture. Moreover, this agreement has been maintained almost from the beginning of the discussion of the nature of agricultural geography. The first sentence of the definition offered in 1964 by Reeds—"Agricultural geography in its broadest sense seeks to describe and explain areal differentiation in agriculture . . ."—has its counterpart in Bernhard's statement of 1915 that "Agricultural geography strives to bring light to the spatial variations in agriculture and the reasons for them." Similarly, Otremba's statement in 1964, that "Area as an object of investigation [by agricultural geography] is understandable only in the perception of the spatial arrangement of its 'material filling,'" has a precursor in Hillman's assertion, in 1911: "Just as we . . . have comparative scientists in other disciplines, like comparative anatomy, etc., so does agricultural geography constitute a systematic comparative agriculture of the countries and continents."[4]

In contrast to this almost complete agreement on the importance of areal variation to agricultural geography are the many disagreements over the context in which these variations are to be studied. Some scholars have felt that such studies should be made principally for discovering and analyzing distributions of economic features; others have preferred to limit the study of economic features to regional complexes. Some have been convinced that studies of areal variation should deal only with the relationships between agricultural activities and the physical environment; others have replied that such a restriction would arbitrarily exclude from consideration those many relationships that have little to do

[2] T. H. Engelbrecht, "Der Standort der Landwirtschaftszweige in Nord-Amerika," *Landwirthschaftliche Jahrbücher,* Vol. 12 (1883), 459–509.

[3] R. Krzymowski, "Die wissenschaftliche Stellung der Landwirtschaftsgeographie," *Fühlings landwirtschaftliche Zeitung,* Vol. 60 (1911), 252–65; Waibel, *Probleme der Landwirtschaftsgeographie* (Breslau: Hirt, 1933).

[4] L. G. Reeds, "Agricultural Geography: Progress and Prospects," *Canadian Geographer,* Vol. 8 (1964), 51; H. Bernhard, "Die Agrargeographie als Wissenschaftliche Disciplin," *Petermanns Mitteilungen,* Vol. 61 (1915), 12 f.; E. Otremba, "Die Landwirtschaftswissenschaften in Beziehung zur Landwirtschaftsgeographie," *Erde,* Vol. 95 (1964), 299; R. Hillman, "Die landwirtschaftliche Erdkunde als Gegenstand des Hochschulunterrichts," *Fühlings landwirtschaftliche Zeitung,* Vol. 60 (1911), 296.

directly with the environment. Some have emphasized the derivation of generalizations, or "laws"; others have maintained that studies of variation should dwell largely on the unique, or at most give equal attention to the two objectives. Even on a point that one might expect to evoke universal approval, the need for an applied aspect of agricultural geography, not all methodological writers have agreed, particularly not on the amount of emphasis to be given to it.

Agriculturists and economists have argued some of these points as well as geographers, so that views on the purposes of agricultural geography are also not infrequently combined with opinions on the position of the field relative to geography, the agricultural sciences, and economics. Early writings by geographers on the objectives of economic—and by implication, agricultural—geography leave no doubt as to what discipline they felt should cultivate the subject. Friedrich viewed the goal of economic geography as one of describing and explaining the "geographic distribution of economic facts as spatial phenomena on the earth surface" in terms of historical development, present situation, and quantity and quality. However, the "gathering of information on the economic situation in the individual world areas without considering the objective of combining the facts with nature and man" would be assigned to economics.[5] Hettner generally agreed with this allocation of subject matter, although he emphasized the study of variations in the "economic character" of regions as the principal goal of economic geography:

> The knowledge of the geographic distribution of individual products or products belongs to the sciences of economic production . . . and can be designated as a geographic *Produktenkunde* [literally, "product information"]; economic geography, in contrast, has to do with the economic characteristics and relationships of the various countries and localities.[6]

Schlüter also defined economic geography by comparing it to economics:

> The political economist directs his view toward the economic processes. The kinds of exchanges, the means by which they are accomplished, thus, above all, money and trade, the causes and effects of the entire process that inserts itself between production and consumption of goods—these are the facts that properly concern him, at the same time always with consideration of the practical results, for better or worse. One can also consider these things in their relation to the earth surface and, for example, investigate the influence that the geographic situation can exert on the battle of competition. But this still brings us only a geographic political economy, which stands in exactly the same relationship to geography as does geographic botany.[7]

[5] E. Friedrich, *Allgemeine und spezielle Wirtschaftsgeographie,* 2d ed. (Leipzig: Göschen, 1907), p. 11.

[6] A. Hettner, "Das Wesen und die Methoden der Geographie," *Geographische Zeitschrift,* Vol. 2 (1905).

[7] O. Schlüter, *Die Ziele der Geographie des Menschen* (Munich, 1906), p. 31.

Several economists, however, have disputed these proprietary views. Harms has held that only within the economic sciences can the work of economic geography be adequately pursued, namely the investigation of the "reciprocal effect of the various natural conditions . . . in their significance for the total meshing of economic activity." Keasbey and Robinson have viewed the subject as a border discipline, bridging the gap between economics and geography. But Keasbey does agree with Harms that economic geography is the "investigation of natural conditions in their effect on human economy," while Robinson sees the need for the field in the understanding of the areal variations in the quality and volume of economic processes.[8]

Bernhard has still another view of the purposes and position of economic geography. He agrees with Keasbey and Robinson on economic geography as being a bridging field but also advocates greater independence for agricultural geography. For him, agricultural geography is "both agricultural science and geography; from the former it takes its object, from the latter its approach." Rejected are the claims of Friedrich, Hettner, and Schlüter that the field belongs only to geography; to reserve *Wirtschaftsgeographie* (economic geography) to geography and *geographische Wirtschaftskunde* (literally, "geographic economic information") to economists is to make a meaningless and methodologically impossible division. Bernhard feels this is particularly true of Friedrich's assertion that workers in *Wirtschaftskunde* (his geographic branch of economics) should concern themselves "only with the gathering of information on the situation of the economy in the individual world areas, without considering the objective of combining the facts with nature and man." For Bernhard, there "is little value in undertaking comparative considerations of economic conditions in various world areas if one does not wish to draw on natural and human factors for the explanation of spatial variations." And to Schlüter and Hettner, who believe that even a consideration of these relationships would still not be economic geography, but rather a "geographical political economy," Bernhard retorts that this division is one between regional and systematic approaches in geography rather than between geography and economics. However, Bernhard also rebuffs the claims of Harms that the economic geography of the geographers should be abandoned in favor of a "social economic geography" within economics. Geographers, not economists, are trained to study the "factors that evoke the spatial deviations in the economy," is his opinion.

It is to the agricultural sciences, however, and not geography, that Bernhard turns to justify his claim for an agricultural geography that is independent even of economic geography. Ironically, he bases his argument on the very points used by Hettner to justify the position of geography among the sciences. Reality, according to Hettner, is simul-

[8] B. Harms, *Volkswirtschaft und Weltwirtschaft* (Jena: Fischer, 1912); L. M. Keasbey, "The Study of Economic Geography," *Publications of the American Economics Association,* Series 3, Vol. 2 (1901), 166–67; E. V. D. Robinson, "Economic Geography: What It Is and What It Is Not," *ibid.,* Series 3, Vol. 10 (1909), 247–57.

taneously a three-dimensional space which must be viewed from three different points of view in order to understand the whole. One of these viewpoints is that of the systematic sciences, which "find their unity in the objective likeness or similarity of the subjects with which they are concerned." Another viewpoint, that of the historical sciences, comprises efforts to "unite a number of things of entirely different systematic categories and obtain their unity through the point of view of the temporal progression of events in time." The third viewpoint, that justifying the role of geography, encompasses the study of "the arrangement of things in space." [9] Bernhard, however, claims that these same viewpoints have to be considered in appreciating agricultural, as well as general, reality, and offers another "system" of agricultural sciences:

- I. Systematic view
  - A. Fields of agricultural technology
    1. Plant-production fields
    2. Animal-production fields
    3. Allied agricultural-technological fields
  - B. Agricultural economics
- II. Historical view: agricultural history
- III. Geographical view: agricultural geography

Thus, he argues, the development of agricultural knowledge should begin with the study of agricultural history, for "it teaches us to understand the evolution of agriculture with its particular phenomena." After that, agricultural geography should show in a picture of world and regional agriculture "how [the subject] is related to the present spatial conditions of agriculture and in what way natural, economic, and cultural factors share in the associated differentiations." Systematic agricultural science, "the third and last link in the chain, always looks to the future; it should orient itself toward the most successful methods of the agricultural production process." Bernhard therefore concludes that agricultural geography provides the agricultural sciences with the "missing link for the creation of a methodological whole." [10]

Not until 1933 did a geographer attempt to define in some detail both the purpose and position of agricultural geography. Leo Waibel saw the spatial variations of agriculture in the context of the "agricultural formation" (*Landwirtschaftsformation*), the last word of the term deriving from plant geography and referring to plant complexes that have a distinctive landscape, or regional, characterization. Waibel here extended the term to any region that carried the stamp of a particular agricultural activity. He contrasted this regional and morphological approach with the more theoretical and utilitarian purposes of the agricultural sciences; and, although he did not specifically answer Bernhard's philosophical arguments for closer relationships of agricultural geography to agriculture,

[9] A. Hettner, *Die Geographie, ihre Geschichte, ihr Wesen und ihre Methoden* (Breslau: Hirt, 1927), pp. 114–17, 123 f.

[10] Bernhard, *op. cit.*, pp. 100–103.

he at least implied that such a comprehensive view lay beyond the interest of the systematically trained scientist.

How comprehensive that view should be was outlined by Waibel in his description of the three "disciplines" of agricultural geography: statistical, ecological, and physiognomic. Statistical agricultural geography uses agricultural statistics to determine the distribution of plants and animals, not only in absolute terms, but ratios, so as to discover the heaviest concentrations of each of these agricultural units, what position each has in the whole farming operation, and where each unit is dislodged by a more profitable one. The statistical approach must be complemented by the ecological, however, which has as its object the study of the different ways in which man combines plants and animals into a farming system within a particular environment. This ecological, or functional, agricultural geography thus has far more geographic significance than that of the statistical. The capstone of the three disciplines is physiognomic agricultural geography. It is not only an attempt at a complete description of the agriculturally conditioned landscape, including crops, roads, paths, farmsteads, and any other material evidences of the farming process; it also seeks to determine the spatial structure and distribution of the various landscapes.[11]

Thirteen, and again sixteen, years later, in 1946 and 1949, Faucher was also to use the comprehensive regional approach as a justification for setting agricultural geography apart from the agricultural sciences. Moreover, he was also to claim it as a basis for subdividing agricultural geography and making one subdivision independent of economic geography. The agriculturist, Faucher said, "researches the technical conditions of production and the means for improving them," whereas the geographer "is more attentive to the results of cultivation than to its processes. The nature of the products, the economic conditions for their obtainment, the way of life of the cultivators, the characteristics and transformations of the rural landscape, constitute his particular object." But this approach, Faucher maintained, should not be confused with the study of world distribution of crops, their volume of production, their use, and their movement. This was the province of the "economic geography of agriculture," a "quantitative" geography based on statistics. The other geographic approach to agriculture was "qualitative" and related more to human than economic geography. Faucher called it "agrarian geography" (*géographie agraire*).[12]

A more liberal view of the relationships between agricultural geography and related disciplines has since been advanced by Otremba. For him, the conceptual impossibility of separating the agricultural landscape from the cultural landscape, and thus denying the visible unity of habitation and farmstead, is the best proof that separation of agricultural geography from the other geographic fields is "highly problematical." The observable trend in the literature of agricultural geography, toward a closer liaison with settlement and social geography, is "only to be wel-

[11] Waibel, *op. cit.*, pp. 8–11.
[12] D. Faucher, *Géographie agraire* (Paris: Médicis, 1949), p. 11.

comed, for it conforms to the worthy goal of the unity of economy and culture in the landscape." But the trend does not encourage Otremba, as it does Faucher, to call for an "agrarian geography autonomous of economic geography." Otremba feels that the overlapping of agricultural geography and economics is just as great as that of agricultural geography and the rest of geography, and that "it is neither wise nor necessary to draw sharp subject boundaries." An approach to the idea that the field is a bridge, espoused earlier by Bernhard and others, can also be noted in Otremba's emphasis on agricultural geography as one of the "typical boundary and correlative sciences," and his belief that "therein lies its special attraction." [13]

Relationships Between Man and His Environment

Not all debates over the nature of agricultural geography have concentrated on both the purpose and position of the field, however. Some objectives have been so commonly accepted as geographic that debate about them has been largely concerned with their interpretation, and little with their justification as a part of geography. Probably the oldest of such arguments has been about man and his relationship to his environment. Many writers, particularly in the early decades of economic geography, felt that the field should concentrate on the significance of the physical environment for human activity. Götz, the first to publish a program for the organization of economic geography (1882), professed to see as its intrinsic goal the "view of earth areas as bases for . . . commercial life." Forty-four years later, in 1926, Passarge stated that economic geography should treat "the dependence of economic phenomena on the quality of earth areas and the transformation of the same through the economy of man." The same year, Whitbeck wrote that the study of relationships with the environment deserved a special title, the "science of geonomics." In the 1930's, one still found opinions favoring the environmental role, such as those of Vallaux, for whom economic geography was the study of human groups adapting to their natural environment, and of Visher, for whom geography's "special field" was the "influence of the natural environment upon the nature and distribution of man's activities and qualities." [14]

Today only a few geographers accept this strongly environmentalist view of economic activity. The record of contrary opinion is also a long one. In 1907, Hettner was already maintaining that with the environ-

[13] E. Otremba, *Allgemeine Agrar- und Industriegeographie* (Stuttgart: Franckh, 1953), pp. 19 f.

[14] W. Götz, "Die Aufgabe der 'wirtschaftlichen Geographie' ('Handelsgeographie')," *Zeitschrift der Gesellschaft für Erdkunde zu Berlin,* Vol. 18 (1882), 364 f; S. Passarge, *Die Erde und ihre Wirtschaftsleben* (Hamburg: Hanseatische Verlagsanstalt, 1926), p. viii, quoted in A. Rühl, *Einführung in die allgemeine Wirtschaftsgeographie* (Leiden: Sijthoff, 1938; reprinted 1968), p. 47; R. H. Whitbeck, "A Science of Geonomics," *Annals of the Association of American Geographers,* Vol. 16 (1926), 117–23; C. Vallaux, "Human Geography and the Social Sciences," *Encyclopedia of the Social Sciences* (New York: Macmillan, 1930–35), Vol. 6 (1931), pp. 624–26; S. S. Visher, "Recent Trends in Geography," *Scientific Monthly,* Vol. 35 (1932), 439.

mentalist approach, "one can usually arrive only at possibilities, for the decision as to whether the natural qualifications actually exert the claimed influence lies not in nature but with man." This idea was again expressed and considerably expanded in the next decade, by Vidal de la Blache, whose work was capped by *Principes de géographie humaine,* posthumously published in 1922. Vidal's statements were further amplified that same year by Fèbvre, who applied the term "possibilism" to the concept. Completing this unusual convergence of thought toward a more humanistic economic geography was a closely related proposal, made by Lütgens in 1921, that the subject be considered "the science of the reciprocal action between the earth area with its contents and economic man, and therewith of the distribution and explanation of its manifestations and results." [15]

The "reciprocal" and "possibilist" concepts also found their way, later, into discussions concentrating on agricultural geography. Buchanan, in the Preface to his 1935 study of the New Zealand livestock industries, reflected Lütgens' viewpoint when he asserted that "the geography of production must be a study of the *interaction* of geographical and economic conditions in the area and for the products concerned, and it is such a study of mutual cause and effect that is here attempted," [16] and Hettner rephrased his views about possibilism in more agricultural-geographic terms, in the first volume of his *Allgemeine Geographie des Menschen,* published posthumously in 1947:

> Nature, in its diverse conditions: in the soil, in the water, in the climate and in the natural existence or the transfer of suitable plants and animals, offers in different earth areas completely different possibilities to which man can not think himself superior; the knowledge of their geographic distribution and the resulting . . . potential agricultural zones and regions is the prerequisite. But whether and how man exploits these possibilities, whether and how far the potential areas gain agricultural reality, depends on his culture and national individuality.[17]

New conceptions have continued to arise, and all involve still greater attention to the human side of the relationship. Thoman expanded the structure of the relationship by writing of the "interactions of man, culture and nature." His is not the "culture" that geographers have commonly considered, i.e., the man-made features of the landscape, but the "cultural legacy upon which . . . minds rely for instruments to do the job." Somewhat the same approach has been offered by Wagner, who

[15] A. Hettner, "Die Geographie des Menschen," *Geographische Zeitschrift,* Vol. 13 (1907), 13; P. V. de la Blache, *Principes de géographie humaine.* Ed. by E. de Martonne (Paris: Armand Colin, 1922); L. Fèbvre, *La terre et l'évolution humaine* (Paris: La Renaissance du Livre, 1922); R. Lütgens, "Spezielle Wirtschaftsgeographie auf landschaftskundlicher Grundlage," *Mitteilungen der geographischen Gesellschaft Hamburg,* Vol. 33 (1921), 136.

[16] R. O. Buchanan, "The Pastoral Industries of New Zealand," *Transactions, Institute of British Geographers,* No. 2 (1935), xv.

[17] A. Hettner, *Allgemeine Geographie des Menschen,* Vol. 1: *Die Menschheit* (Stuttgart: Kohlhammer, 1947), pp. 6, 306.

observes that although differences of natural environment play roles in the operation of "technical systems" such as crop rotation or land tenure, still another kind of environment requires consideration, society: "The societal environment has, like the natural environment, great significance for technical processes." More attention to the "social environment" is also asked by Gregor, who complains that "although most economic—and human—geography books commonly emphasize the visible works of man, they ignore many of the genuinely cultural reasons for those accomplishments." More recently, Otremba has referred to a "triangular system" of "man—economy—nature," in which "economy" is termed a function of society.[18]

This growing concern with the more cultural aspects of man's relationship to his environment has also found its way into methodological statements specifically on agricultural geography. Here, according to George, one must pay particular attention to three basic sets of relationships: those between the physical environment and agricultural operations; those between population density and the available agricultural space; and an ensemble of historical relationships. The study of the first set is basic in agricultural geography, because "the distinguishing of agricultural domains, defined by their natural aptitude for certain farming operations and by their inaptitude for certain others, is an indispensable basis for geographic discrimination." George cautions, however, against any tendency to overemphasize the force of physical barriers to agriculture by noting that the relations between farming and the environment present "a varied gamut of actions and reactions, of which the lines of force differ according to the technological capacities and the faculties for organization of the agricultural groups." Moreover, George adds, the concept of the absolute barrier to agriculture, the desert, must be overhauled. The numerous successful agricultural implantations made in these sterile lands, whenever economic or strategic needs have become great enough, make it necessary to think of deserts in more relative terms. The majority of them "are defined by the presence of an inhibitive factor associated with natural, neutral, or favorable conditions."

The second category of relationships, that between population density and available agricultural space, is directly conditioned by the intensity of land use, the choice of crops, farming methods, the relationship between cropping and livestock raising, and the forms and localization of the habitat. Variations in population therefore involve modifications of agriculture. The third category, historical relationships, George believes to be important because of the conservative milieu of rural areas. Habits acquired in a previous period tend to survive the conditions that originally favored them. Cereal grains, for example, still play a prominent part in the western European diet although its over-all character has changed

[18] R. S. Thoman, *The Geography of Economic Activity* (New York: McGraw-Hill, 1962), p. 16; P. L. Wagner, *The Human Use of the Earth* (New York: The Free Press, 1964), p. 7; H. F. Gregor, *Environment and Economic Life* (Princeton, N.J.: Van Nostrand, 1963), p. vii; E. Otremba, "Gedanken zur gegenwärtigen Lage der Wirtschaftsgeographie, anhand einer frühen Arbeit von Gottfried Pfeifer," *Geographische Zeitschrift,* Vol. 54 (1966), 9.

considerably. Wheat has also been spread over a large part of the world by the European during his nineteenth-century commercial and political expansion. Thus what were largely climatic facts in the beginning have become historical facts; and these changing relationships between diet and its supporting agricultural forms, as well as the resulting spatial relationships, become "one of the groups of fundamental relationships in agricultural geography." [19]

Yet one must note that George was less concerned with the nature of these relationships than he was with what they expressed. That many of them had little causal association with the environment was immaterial; that they all combined into significant areal associations, in the forms of "rural landscapes" (*paysages rurales*), was important. Relationships were still a necessary object of study, but only insofar as they formed an areal structure.

A similar but even more pronounced view has been advanced by Otremba. Like George, he considers all kinds of relationships and gives much more attention to those of an historical, economic, and social character than he does to those with the natural environment; but these relationships are more implied than declared, their contributions being expressed only in terms of the nature and development of the various "form elements" (*Gestaltelemente*) of the "agricultural landscapes" (*Agrarlandschaften*). This emphasis on form and landscape does not connote a restriction of the field of study to visible features, however, any more than it was connoted by George's *paysages.* From the beginning, Otremba is concerned with viewing the "agriculturally transformed" area "as a whole as well as in its parts, in its outer appearance, in its inner structure and in its interweaving." [20]

Neither Otremba nor George are saying anything new when they affirm the study of relationships as a means of achieving a still higher goal, the delimitation and comprehension of agricultural regions. Most of the methodological writers who have stressed the importance of relationships have done so with an eye on regions. Thus Hettner, although he agreed that economic geography is the "science of reciprocal effects," also declared that "its truer object is not clearly denoted by that." Rather, economic geography seeks to comprehend the "total character of economic life in different areas," which comprises a "definite range of facts, which must first be described and then causally explained." [21] Vidal spoke of the "geography of the whole" as the "highest goal of geographic study." Lütgens immediately followed his statement on "reciprocal action" with a reference to the "economic landscape" as the "final goal." Some agricultural geographers, like Otremba, have been even more emphatic. Waibel, as we have seen, considered the "agricultural formation" as the basic structure for agricultural-geographic study, as did Cholley some-

[19] P. George, *La campagne* (Paris: Presses univ. de France, 1956), pp. 7–9.

[20] E. Otremba, *Allgemeine Agrar- und Industriegeographie,* 2d ed. rev. (Stuttgart: Franckh, 1960), p. 25.

[21] A. Hettner, *Allgemeine Geographie des Menschen,* Vol. 2, *Wirtschaftsgeographie* (Stuttgart: Kohlhammer, 1957), p. 13.

what later when he extolled the "agrarian structure." [22] Buchanan, in his most intensive methodological study of the field, devoted practically all of his time to problems of delimiting agricultural regions. That Waibel and Cholley, however, viewed the agricultural region as a complete physical entity while Buchanan saw it only as a spatial reflection of a farming system shows that agreement on the primacy of the region did not always extend to its definition.

Theoretical Geography

More recently, a small but growing group, principally American geographers, has challenged the traditional esteem of methodologists for regions. They believe that the only way that geographers can contribute meaningfully to the comprehension of areal variation is to develop theories or laws that will account for locations of economic activities. Since the derivation of universals depends on dealing with only a small number of independent variables, all subject to similar laws, the unsuitability of the regional approach for this research method is obvious. Moreover, many of these variables are to be found beyond, as well as within, the locational complex. This is especially so when one considers only the functional, or "systems," aspect of economic activity and excludes the landscape, or "facilities," aspect, as do the theoretical geographers.

The complications introduced by variables have also led the theoretical geographers to emphasize association rather than cause in explaining locational patterns, in contrast to the "environmentalists," who also have stressed location, and even, in part, to the "regionalists." By thinking in terms of association rather than cause and effect, the theoretical geographer hopes to reduce at least partially the chances of being confronted by too many variables, a challenge readily accepted by those who search for causes, but an impediment to those seeking solutions to locational problems. The frequent use of mathematical terms by theoretical geographers is still another illustration of their concern with a more scientific approach to locational problems. This concern also finds outlet in advocacies of greater quantification in the description of spatial associations.

It is not surprising that many of the first geographers who expressed a wish for a more scientific geography were also men who had done their major work in agricultural geography. What is surprising is their prescience, for all of them spoke well before the rise of the so-called theoretical movement. Bowman's comments on "Commercial Geography as a Science" in 1925 are perhaps the outstanding example. Economic geography books were, according to Bowman, "loaded with inappropriate descriptive detail," but he also thought that this would only be a "passing stage," to be followed shortly by "a new organization of principles or

[22] A. Cholley, "Problèmes de structure agraire et d'économie rurale," *Annales de Géographie,* Vol. 60 (1946), 81–101.

laws." In his summary of those conditions that he felt economic geography must fulfill in order "to be a science," Bowman wrote:

> [I]t should (1) have its facts arranged in a systematic order; (2) rationally explain present conditions in accordance with established "laws"; (3) predict the future course of development or lead the way to the discovery of new laws. These it should do not in any absolute or infallible sense but with some degree of accuracy. The tests applied by one person in a given solution must be capable of verification at other places and by other men under standard conditions, just as a physicist or a physiologist lays down the governing conditions of his experiment and confidently invites verification of a truth or law which he claims to have discovered. The terms of discussion are in the main specific, not general, and every step in a proof must rest upon facts and a logical progression of argument.[23]

Ten years after Bowman's remarks, Colby wrote that "The application of statistical methods, I believe, will introduce new types of measurement, will clarify present methods of analysis, and will give us results which are quantitatively exact as well as qualitatively true." [24]

Forethoughts such as these were not confined to American geographers. Rühl was already describing economic geography as a "differential location science" in 1918, and in the early 1930's was calling for a "theoretical" economic geography, one that would seek "to discover the general rules for the distribution of products, investigate economic forms, and develop the concepts and terminology necessary for scientific comprehension." It was also in the Thirties that Waibel discoursed at length on the significance for agricultural geography of Thünen's "law," listing as the final objective of agricultural geography the investigation of the "spatial arrangement and distribution" of agricultural regions.[25]

Opponents of the theoretical school are skeptical of the extent to which phenomena in human geography can provide similar cases and, therefore, material for the formation of laws. Hartshorne observes that even the study of the relationships of domesticated plants and animals is complicated by the varieties that man has developed, and continues to develop, from the same species; and as the phenomena studied become more dependent on man, the great differences in technology, economic levels, and institutions over the earth force one to increasingly narrow the area of study if one is to establish generic concepts and principles. While a study of the production of rice in China might well produce such concepts and principles, they could not be assumed to be applicable outside of the country. In this connection, Robinson also questions the use of the word "theory" by the theoretical geographers, noting that "when one is dealing with physical laws, theories are basic explanations or descriptions of

[23] I. Bowman, "Commercial Geography as a Science," *Geographical Review*, Vol. 15 (1925), 288 f.

[24] C. C. Colby, "Changing Currents of Geographic Thought in America," *Annals of the Association of American Geographers*, Vol. 26 (1936), 37.

[25] Rühl, "Aufgaben und Stellung der Wirtschaftsgeographie," *Zeitschrift der Gesellschaft für Erdkunde zu Berlin*, Vol. 54 (1918), 500; *idem*, *Einführung in die allgemeine Wirtschaftsgeographie*, p. 64; Waibel, *op. cit.*, p. 11.

their interaction that, if correct, will apply without regard to time or location."[26]

A correlate of this objection to an all-embracing theoretical geography is the criticism that theoretical geographers tend to undertake problems only if they can be attacked by quantitative methods. Given enough time for the further development of these methods and the education of a larger number of researchers in the application of these techniques, such a restricted approach, it is argued, would inevitably lead to a serious substantive bias in geographical research. Nor are all theoretical geographers in complete agreement on their approach. McCarty, for example, feels that the possibility of evolving a more general body of theory is greatest when one works from small areas to larger ones, while Warntz equates theoretical economic geography with "macroeconomic" geography and traditional economic geography with "microeconomic" geography.[27]

Replies by theoretical geographers have not been lacking. It is also worth noting that one of the most direct answers comes in connection with agricultural geography. To the charge that searching for laws among human variables is an impossible task, Harvey notes that this assertion ignores the tremendous strides that have been made in the behavioral sciences and the resulting prospect of building models of agricultural systems which take account of sociological and psychological realities. Such models will provide the understanding of decision-making processes which is needed, he says, because "Land-use patterns are, after all, the end product (or geographical expression) of a large number of individual decisions made at different times for often very different reasons (or perhaps for no adequate reason at all)." Coincidentally, one of the sharpest critics of these models is another researcher interested in the geographical aspects of agriculture, in this case the location of agricultural production. Models, according to Dunn, an economist, are incapable of direct application to the solution of locational problems because "there are too many sources of discontinuity and elements of indeterminacy in the realistic case." Even if these difficulties could be solved, Dunn continues, the expense and time required to obtain the basic data would make a general solution impractical.[28]

Pure vs. Practical Objectives

Debates about the pertinence of the theoretical approach to agricultural geography have their counterpart in the arguments over the relative

[26] R. Hartshorne, *Perspective on the Nature of Geography* (Chicago: Rand McNally, 1959), p. 150; A. H. Robinson, "On Perks and Pokes," *Economic Geography,* Vol. 37 (1961), 182.

[27] H. H. McCarty, "An Approach to a Theory of Economic Geography," *ibid.,* Vol. 30 (1954), p. 96; W. Warntz, "Progress in Economic Geography," in *New Viewpoints in Geography,* ed. P. E. James (Baltimore: National Council for the Social Studies, 1959), Vol. 5, Part 1, pp. 74–75.

[28] D. W. Harvey, "Theoretical Concepts and the Analysis of Agricultural Land Use Patterns in Geography," *Annals of the Association of American Geographers,* Vol. 56 (1966), 370; E. S. Dunn, Jr., *The Location of Agricultural Production* (Gainesville: Univ. of Florida, 1954), p. 93.

importance of "pure" and practical objectives. These arguments are almost as old as the methodological discussion in agricultural geography, despite the innately practical character of the field and the attention given to its practical aspects. Nor do we find any trend in the literature, nor evidence that the particular training of an author affects his position. One of the very few appeals for a pure agricultural geography was made when the field was still hardly recognized as an entity, and by Bernhard, a researcher trained in the applied aspects of agriculture. He believed in a geographic approach, "even if its accomplishment is capable of giving no prospect of practical benefit." His goal rested upon the assurance that "a scientific fact deserves to be sought and recognized for itself," although the facts were nevertheless to be used: "agricultural geography is at the service of both agriculture and geography." Twenty-three years later, in 1938, Bernhard received support from geographer Rühl's discussion of economic geography. To Rühl, devoting time to social objectives would make vain the "years-long battle" for the objectivity of information in the social sciences. These sciences, as all others, "serve life best when they seem to be most distant from it." [29]

Equal attention to pure and practical objectives has also been advocated, first by Götz in his proposals of 1882 for a field of economic geography, and fifty years later in Barrows' side reference to "economic regional geography." Barrows believed that this field faced "two basic problems . . . closely related to each other but yet distant: (1) how does man use the land and its resources, and why does he use them as he does? and (2) what are the advantages and disadvantages, the opportunities and handicaps, of the region for utilization by man?" Colby, in 1936, especially emphasized the complementarity of these two centers of geographic interest, the "philosophical and scientific" and the "social and practical." For him, consideration of areas "inevitably" led to consideration of uses of the land and other resources, and problems having to do with resources could not be understood or solved unless they were properly placed in their areal setting. Much the same concern for a balanced view of pure and practical objectives was evinced during the closing years of the last world war by Finch and Ackerman, who felt that geographers had favored the more academic aspects of geography to the slighting of research on national and international problems.[30]

Agriculturist and geographer contribute to still another shade of opinion on the relative importance of pure and practical objectives in research in agricultural geography: the favoring of both types of goals, but with the assumption that pure research is preliminary to practical. Krzymowski, an agricultural specialist who in 1911 preceded Bernhard

[29] Bernhard, *op. cit.*, pp. 13, 103; Rühl, *Einführung in die allgemeine Wirtschaftsgeographie*, p. 65.

[30] Götz, *op. cit.*, pp. 354–88; H. H. Barrows, "Geography as Human Ecology," *Annals of the Association of American Geographers*, Vol. 13 (1923), 9; Colby, *op. cit.*, p. 29; V. C. Finch, "Training for Research in Economic Geography," *Annals of the Association of American Geographers*, Vol. 34 (1944), 207–15; E. A. Ackerman, "Geographic Training, Wartime Research, and Immediate Professional Objectives," *ibid.*, Vol. 35 (1945), 121–43.

by four years in his advocacy of a field of agricultural geography, based his argument on the need for providing agriculturists with all the information about how farmers have adjusted their farming operations to different physical and economic situations. Although the goal of researchers collecting this information was "pure knowledge in and for itself," the data was eventually to furnish raw material for theoretical speculation by the agriculturist. Hillmann, a contemporary, advanced essentially the same argument in support of his own contention that agricultural geography should be included in the university curriculum.[31]

Half a century later, McCarty and Lindberg seemed to agree with Hillmann and Krzymowski, at least in principle. The work of the economic geographer, they wrote, consists of "finding solutions for problems . . . that begin with relatively simple questions such as, 'Why is this phenomenon located here?' and proceed to more complex queries such as, 'Where can I most advantageously locate my business?' and 'What types of economic development would be most successful in this area?' " Further, rigorous standards of research "must prevail if the discipline of geography is to be of maximum service to mankind." A similar partiality for the practical side, but with an emphasis on its difference from the academic concept rather than on any organic relationship, has recently been offered by Symons in a statement on agricultural geography. He notes that although the research worker faces "endless problems of an academic nature," he also is confronted by the "greater problems that beset mankind as a whole." Any contribution that agricultural geography can make to the solving of the greatest of these problems—hunger—is, Symons asserts, "ultimately more important than its academic justification." [32]

The extreme of the view that the practical orientation of agricultural geography is primary is held by such men as du Plessis de Grenédan, Baker, and Bowman. For them, the subject was strictly utilitarian; without a deliberate effort to contribute to social welfare, one could not justify his efforts in the field. Grenédan, an agriculturist, advocated this approach in 1903, in what is probably the first modern text on agricultural geography. Since the welfare of the farmer depends on many things besides physical conditions, such as market conditions and transport programs, the agricultural geography of Grenédan was far more inclusive than many geographers admit.[33] Several decades later, Baker, the pioneer in American agricultural geography, affirmed that the geography of the future should offer information that would enable the farmer, miner, engineer, and industrialist to use natural resources to the greatest possible economic advantage. But it was Bowman who, in justifying his major work on frontier agricultural settlement, made the most vigorous assertions of all:

[31] Krzymowski, *op. cit.*, pp. 258, 264; Hillmann, *op. cit.*, pp. 289–97.

[32] H. H. McCarty and J. B. Lindberg, *A Preface to Economic Geography* (Englewood Cliffs, N.J.: Prentice-Hall, 1966), pp. v f; L. Symons, *Agricultural Geography* (New York: Praeger, 1967), p. 258.

[33] J. du Plessis de Grenédan, *Géographie agricole de la France et du monde* (Paris: Masson, 1903), pp. 2 f.

Most of the papers dealing with land-use seem to be more concerned with the development of techniques than with ideas. . . . If geography is ever to influence political and social policies it must deal with ideas that seem to be of critical importance to government and society, conveyed in terms that leaders can understand.[34]

Thus, advocates of a pure agricultural geography, like advocates of an unqualified practical approach, can be found among both geographers and agriculturists and in both earlier and more recent periods. Opinion about how far the geographer can go with the practical approach and still remain within the competence of his field is less divided and more cautious, however. Symptomatic is George's distinction between a *géographie active* and a *géographie appliqué*. The mission of the first is to "define the needs, possibilities, and options" of life in a region; that of the second is the "application of information furnished by geography, which is the business of administrators." Despite this more literal use of the word "applied," Colby earlier came to the same conclusion in effect when he drew a line between the "design" and "effectuation" phases of the "planning process." Likewise, Stamp, probably the foremost exponent of geographers in land use planning, felt that although the "subjective judgement" of the geographer "is certainly deserving of full consideration," the geographer could "consider his work ends with survey and analysis." The most striking example of this cautious attitude toward the practical role of the geographer is found among Soviet methodologists, from a land where practical goals are uppermost in geography. A recent pronouncement, Konstantinov's, warns economic geographers to concentrate on "pre-planning" and "pre-design" work, and not to try and replace staff members of planning and design agencies. To be sure, his is a recommendation based more on the desire to maintain a high level of specialization, and therefore efficiency, than on a wish to keep geography "pure."[35]

Methodological and Subject-Matter Research

Although opinions on the nature of agricultural geography indicate general trends and orientations of research, they still provide us with only the philosophical considerations involved in the field. Remaining to be considered is the incomparably larger body of contributions to the subject matter of the field. Indeed, many would argue that these materials are even better indicators of what geographers think agricultural geography is. We shall now proceed to note some of the more important advances made in the techniques used in the varied research quests of those interested in agricultural geography.

[34] I. Bowman, "Planning in Pioneer Settlement," *Annals of the Association of American Geographers,* Vol. 22 (1932), 93.

[35] P. George, *La géographie active* (Paris: Presses univ. de France, 1964), pp. 35, 37; Colby, *op. cit.,* p. 34; L. D. Stamp, *Applied Geography* (Baltimore: Penguin, 1960), p. 19; O. A. Konstantinov, "On the Thirtieth Anniversary of the Department of Economic Geography of the Geographical Society USSR," *Soviet Geography: Review and Translation,* Vol. 7 (1965), No. 7, 9.

CHAPTER TWO

Research methods

Field Work

Systematic field work has long played a particularly prominent role in agricultural geography. It was already being emphasized in this country as early as 1915—well before economic geography had become seriously established in the United States—by Jones and Sauer in their detailed "Outline for Field Work in Geography." This was followed, nine years later, by a methodological justification by Sauer, in which he posed the possibility that field work might one day become as important to geography as it had already proved for geology.[1] Supporting this thesis was the conclusion by a group of prominent geographers that "sound generalizations about a region should be based on intensive studies of typical small areas."[2] Again, as Jones and Sauer had earlier, the participants focused their attention on the kinds of details to be mapped. Soon after, Whittlesey tested their conclusions in a field study of a six-and-a-half square mile area in eastern Wisconsin, mapping both "natural" and "cultural" features of the environment at a scale of six inches to the mile.[3] With the work of these authors, the detailed mapping of small areas quickly became a major theme of American agricultural geography.

American geographers have also been particularly concerned about the great amount of time needed for mapping, for even small areas can

[1] W. D. Jones and C. O. Sauer, "Outline for Field Work in Geography," *Bulletin of the American Geographical Society*, Vol. 47 (1915), 520–25; C. O. Sauer, "The Survey Method in Geography and Its Objectives," *Annals of the Association of American Geographers*, Vol. 14 (1924), 20.

[2] W. D. Jones and V. C. Finch, "Detailed Field Mapping in the Study of the Economic Geography of an Agricultural Area," *ibid.*, Vol. 15 (1925), 148–57.

[3] D. S. Whittlesey, "Field Maps for the Geography of an Agricultural Area," *ibid.*, Vol. 15 (1925), 187–91.

have a wealth of land use detail. One solution, perhaps the most distinctive of American contributions to field techniques, is the "fractional code" system of mapping. It is a shorthand method of classifying areal associations. Each item is represented by a digit, those features referring to types of land use being grouped in the numerator, and the remainder, denoting natural physical conditions, being assigned to the denominator. First described by Jones in 1930, the fractional code system was later developed and used in an intensive field study by Finch who concluded that it was still too time-consuming. In 1936, under the direction of Hudson, geographers working for the Tennessee Valley Administration reduced the mapping time considerably by using aerial photographs as bases for plotting field data.[4]

Although the fractional code has proved more helpful to American geographers than to Europeans, because of the less detailed map coverage of North America, it remained for Credner, a German geographer, to demonstrate what the method can do in a full characterization of an agricultural region. The cultural aspects of the region—historical development, field form, land inheritance system, degree of urban influence—comprise his "numerator elements" (*Zählerelemente*); the economic aspects of the region—economic form, farming system, rotation system, principal crops—form the "denominator elements" (*Nennerelemente*). More recently, the fractional code was used in an extensive mapping project, surveys conducted by Canadian government geographers of numerous sample areas of their country, and in the Rural Land Classification Program completed for Puerto Rico by the Geography Department of Northwestern University.[5]

Still another shortcut in field work is sampling, the study of selected parts of the survey area. Sampling is done either by "working" traverses or selecting a pattern of dispersed areas. Proudfoot has outlined the basic procedures of the traverse method, although the shape and length can be highly varied. Thus, although most traverses are best accomplished in a straight line, Colby has demonstrated the value of a more sinuous one where large coverage is desired in a short time and where the principal transportation is the railroad. Lengths of traverses in most studies have been only a few miles long, but where only a very limited part of the rural landscape is to be recorded, distances can be much greater, as Finley and Scott demonstrated beautifully in their "Great Lakes-to-Gulf

[4] W. D. Jones, "A Method of Determining the Degree of Coincidence in Distribution of Agricultural Uses of Land with Slope-Soil-Drainage Complexes," *Transactions of the Illinois State Academy of Sciences,* Vol. 22 (1930), 549–54; V. C. Finch, *Montfort: A Study in Landscape Types in Southwestern Wisconsin,* Geographical Society of Chicago, Bulletin No. 9 (1933); G. D. Hudson, "The Unit-Area Method of Land Classification," *Annals of the Association of American Geographers,* Vol. 26 (1936), 99–112.

[5] Credner's manuscript was published posthumously in E. Otremba, *Allgemeine Agrar- und Industriegeographie,* 2d ed. (Stuttgart: Franckh, 1960), pp. 52–53, 55–56; J. W. Watson, "Land Use Surveys in Canada," *Proceedings, Eighth General Assembly and Seventeenth International Congress* (Washington, D.C.: National Academy of Sciences, National Research Council, 1952), pp. 181–91; Department of Geography, Northwestern University, *The Rural Land Classification Program of Puerto Rico* (Evanston, Ill.: Northwestern Univ., 1952).

Profile" of farmhouse types. Mather and Hart, in their study of fences, compromised by selecting sixteen four-mile traverses, evenly spaced along a 1,650-mile round-trip route between Cleveland, Ohio, and Athens, Georgia.

Because the dispersed-area method of sampling has been used so frequently in North America, where roads often reflect rectangular cadastral patterns, sample patterns have been commonly of geometric outline. Among the more recent sampling studies, townships have been used intensively by Reeds in his presentation of type-of-farming areas of southern Ontario, and by Ehlers in his study of the different zones of development within the Peace River pioneer fringe.[6]

The positioning of the sample areas has generally been done in one of three ways, none of which meets with the full approval of geographers. The simplest method is systematic dispersal over the entire area, as in Birch's survey of the Isle of Man, for which he selected samples from the frames provided by British Ordnance Survey maps. Another way is to scatter the samples as much as possible, at the same time modifying the geometric spread to ensure that irregularly distributed phenomena judged to be significant by the investigator may also be recorded. This procedure was found especially profitable by the author in a sampling of southern California farmsteads. Bunge and Blaut have called for greater use of random sampling, a method that sacrifices area for accuracy, although Steiner claims that systematic sampling is even more accurate. At least a partial answer to the lack of systematic coverage by the random sample, however, does seem to lie in the "stratified" version used by Wood.[7]

Sample studies are usually supplemented by questionnaires and, whenever possible, by aerial photography. Progress in expanding and improving these aids has been spectacular, occasionally to the point where blanket coverage has become possible and field work by the chief investigator unnecessary. The most successful use of questionnaires was by Dudley Stamp and the British Land Survey Office. They circulated 22,000 field maps, six inches to the mile, among English school children, and asked them to draw in the land use patterns of their locality. The

[6] M. J. Proudfoot, "Sampling with Traverse Lines," *Journal of the American Statistical Association,* Vol. 37 (1942), 265–70; C. C. Colby, "The Railway Traverse as an Aid in Reconnaissance," *Annals of the Association of American Geographers,* Vol. 23 (1933), 137–64; R. Finley and E. M. Scott, "A Great Lakes-to-Gulf Profile of Dispersed Dwelling Types," *Geographical Review,* Vol. 30 (1940), 412–19; E. C. Mather and J. F. Hart, "Fences and Farms," *ibid.,* Vol. 44 (1954), 201–23; L. G. Reeds, "Agricultural Geography: Progress and Prospects," *Canadian Geographer,* Vol. 8 (1964), 55; E. Ehlers, *Das nördliche Peace River Country: Alberta, Kanada* (Tübingen: Univ. Tübingen, 1965), pp. 99 ff.

[7] J. W. Birch, "A Note on the Sample-Farm Survey and Its Uses as a Basis for Generalized Mapping," *Economic Geography,* Vol. 36 (1960), 255; H. F. Gregor, "A Sample Study of the California Ranch," *Annals of the Association of American Geographers,* Vol. 41 (1951), 286; W. B. Bunge, *Theoretical Geography* (Lund: Univ. of Lund, 1962), pp. 97–103; J. M. Blaut, "Microgeographic Sampling: A Quantitative Approach to Regional Agricultural Geography," *Economic Geography,* Vol. 35 (1959), 79–88; R. Steiner, "Some Problems in Designing Samples of Rural Land Use," *Yearbook of the Association of Pacific Coast Geographers,* Vol. 19 (1957), 25; W. F. Wood, "The Use of Stratified Random Samples in a Land Use Study," *Annals of the Association of American Geographers,* Vol. 45 (1955), 350–67.

results, edited and reduced to the scale of one inch to the mile, were subsequently published as colored sheets. The success of the venture encouraged Trewartha to use the same method for the first nationwide comparison of American farmsteads.[8]

The increase in the use and the potentialities of aerial photography has been even more impressive. Aerial photographs were already being used by geographers in individual field work in the United States by the late 1920's. By 1936, as we noted earlier, government geographers had used them in mapping the Tennessee River Valley, probably the largest project of its kind at the time. During and since World War II, aerial photo coverage has been tremendously enlarged by the activities of the military services and other government agencies. European geographers have particularly benefited, for they have been able to use the photos to decipher present and past field patterns as they attempt to explain the long and complex development of their agricultural landscape. American geographers have used the material primarily for the study of current situations. The contrast between the more historical orientation of agricultural-geographic research in Europe and the less genetically inclined work of American geographers is illustrated in two of the more recent attempts to develop photo interpretation keys, one emphasizing the interpretation of archeological sites as indicators of ancient fields in Britain, and the other centering on present combinations of farmsteads and agricultural systems as indicators of types of farming in the United States.[9]

Two of the latest developments in aerial reconnaissance give promise of perhaps as great an advance in field analysis methods as was provided by aerial photography itself. Photographs of infrared radiation are able to significantly heighten tonal contrasts between crops, and between wooded and unwooded areas of uncultivated lands, and Side-Looking Airborne Radar (SLAR) can increase the visual scope of the camera enormously and still produce a readable image of agricultural land use patterns of uniform scale.[10] These and other developing techniques promise even more, however, when photographs taken from satellites become more common.

Advances in archival research, the inseparable companion of field observation, have also greatly facilitated research in agricultural geography. Rapidly accumulating census information has particularly benefited field work, as is well shown in Birch's extensive use of the annual British Agricultural Returns in his survey of the Isle of Man. Forward and

[8] L. D. Stamp, "The Land Utilization Survey of Britain," *Geographical Journal*, Vol. 78 (1931), 40–47; G. T. Trewartha, "Some Regional Characteristics of American Farmsteads," *Annals of the Association of American Geographers*, Vol. 38 (1948), 169–225.

[9] R. M. Newcomb, "Two Keys for the Historical Interpretation of Aerial Photographs," *California Geographer*, Vol. 7 (1966), 37–46; C. F. Kohn, "The Use of Aerial Photographs in the Geographical Analysis of Rural Settlements," *Photogrammetric Engineering*, Vol. 17 (1951), 759–71.

[10] C. E. Olson, Jr., "Accuracy of Land-Use Interpretation from Infrared Imagery in the 4.5 to 5.5 Micron Band," *Annals of the Association of American Geographers*, Vol. 57 (1967), 382–88; R. B. Simpson, "Radar, Geographic Tool," *ibid.*, Vol. 56 (1966), 80–96.

Raymond, on the other hand, have developed a system of generalization from Canadian census statistics that makes it possible to construct small-scale maps of land use without field mapping and aerial photography.[11]

Many more of the studies made possible by these expanded sources of agricultural statistics have been conducted with little thought of their significance for field techniques, either one way or another. The object of these has been to use the data for distinguishing far-flung spatial patterns, whether they be of systems of agricultural regions or of only one or two isolated agricultural elements. Research of this kind especially characterizes American agricultural geography. The first detailed regionalization of American agriculture, worked out by Baker on the basis of census materials, was also one of the first extensive research efforts in American agricultural geography. Among the more recent research efforts based on extensive use of census material, those of Prunty and Weaver have been especially numerous. This emphasis on research utilizing census materials is undoubtedly in good part a reaction to the traditional wealth of material provided by the Census Bureau and to the variety of topics that are constantly being added to the list of items to be reported. A negative encouragement has been the lack of a long historical past in American agriculture, possession of which would have attracted more geographers to archival material less statistically oriented. To be sure, this does not mean that longer agricultural histories or less rich censuses have discouraged high-quality research based on censuses elsewhere. Coppock's work on English farming systems, and Bhatia's, on Indian crop distributions, are two examples.[12]

Geographers and other scholars working in agricultural geography have not been content with raw statistical data. As the material has multiplied in amount and variety and areal coverage has spread, the desire to draw generalizations from it has increased. In this search for uniform criteria, researchers have developed several methods of data manipulation, by far the oldest and most popular of which has been the extraction of ratios between two significant groups of criteria. Engelbrecht was the first to use ratios extensively and his practice of relating one category to the next higher category, such as wheat to grain area or grain area to crop-

[11] J. W. Birch, "Observations on the Delimitation of Farming-Type Regions, with Special Reference to the Isle of Man," *Transactions, Institute of British Geographers,* No. 20 (1954), 141–48; C. N. Forward and C. W. Raymond, "Small-Scale Land Use Mapping from Statistical Data," *Economic Geography,* Vol. 35 (1959), 315–21.

[12] O. E. Baker's study, "Agricultural Regions of North America," appeared in *ibid.,* Vol. 2 (1926), 459–93; Vol. 3 (1927), 50–86, 309–39, 447–65; Vol. 4 (1928), 44–73, 399–433; Vol. 5 (1929), 36–39; Vol. 6 (1930), 166–90, 278–308; Vol. 7 (1931), 109–53, 325–64; Vol. 8 (1932), 325–77; and Vol. 9 (1933), 167–97. M. C. Prunty's articles on crop changes in the South appeared *ibid.,* Vol. 26 (1950), 301–14 and Vol. 27 (1951), 189–208; and in *Geographical Review,* Vol. 42 (1954), 439–61. J. C. Weaver's writings on crop combinations in the Middle West were published *ibid.,* Vol. 44 (1954), 175–200, 560–72; and in *Economic Geography,* Vol. 32 (1956), 237–59.

J. T. Coppock, "Crop, Livestock, and Enterprise Combinations in England and Wales," *ibid.,* Vol. 40 (1964), 65–81; S. S. Bhatia, "Patterns of Crop Concentration and Diversification in India," *ibid.,* Vol. 41 (1965), 39–56.

land, is still the most common method of ratio derivation. Later, Baker and Jones used the ratio in American agricultural geography, emphasizing its use in distinguishing different types of agricultural regions. Subsequently, with the Census Bureau beginning to provide monetary information on production and expenditures, interest in ascertaining the input-output ratio has begun to quicken. The resulting figure, or "value added by farming," was later related to crop acreage to obtain a partial picture of the distribution of production efficiency among the principal agricultural regions of California. Most recently, the emphasis was shifted from land to farm and from state to nation by relating this added value to American farms. Little work has been done in applying ratios to the social aspects of agriculture, although Hartshorne has related farm population to various aspects of agricultural land.[13]

The new opportunities for research provided by the expanded archival material have also increased our awareness of the gaps that still exist in the data. Agricultural geographers in the United States have found it impossible to extend their monetary-based generalizations on American agriculture to the rest of the world because the majority of farmers operate on a full or at least partial subsistence basis. Geographers and agricultural economists from other countries have fallen back on indices such as "man-days," "grain equivalents," or acre-yields, but most of these units have been applied only to limited areas.[14] Frequent complaints have also been made about the lack of correspondence between the areas used by census authorities and those that interest the geographer. Even in the Middle West, where a dense net of small counties of fairly equal size should presumably reflect most important spatial variations of data, Weaver has found that the smaller township would have better suited his purposes. In the West, where counties are larger and frequently include or cut across combinations of oases and wasteland, the author found the situation for the geographer even more frustrating.[15]

Census definitions and statistical groupings also pose problems. Especially vexing, as Prunty points out, is the common census definition of a

[13] T. H. Engelbrecht, "Der Standort der Landwirtschaftszweige in Nord-Amerika," *Landwirtschaftliche Jahrbücher,* Vol. 12 (1883), 459–509; Baker, *op. cit.;* W. D. Jones, "Ratios and Isopleth Maps in Regional Investigation of Agricultural Land Occupance," *Annals of the Association of American Geographers,* Vol. 20 (1930), 177–95.

The "value added to farming" was related to crop acreage by H. F. Gregor, "Regional Hierarchies in California Agricultural Production: 1939–1954," *ibid.,* Vol. 53 (1963), 27–37. For the relation of "value added" to farms, see C. W. Olmstead and V. P. Manley, "The Geography of Input, Output, and Scale of Operation in American Agriculture," in *Agricultural Geography: I.G.U. Symposium,* ed. E. S. Simpson (Liverpool: Univ. of Liverpool, 1965), pp. 37–47. R. Hartshorne, "Agricultural Land in Proportion to Agricultural Population in the United States," *Geographical Review,* Vol. 29 (1939), 488–92.

[14] S. S. Bhatia, "A New Measure of Agricultural Efficiency in India," *Economic Geography,* Vol. 43 (1967), 244–60; B. Andreae, *Betriebsformen in der Landwirtschaft* (Stuttgart: Ulmer, 1964).

[15] J. C. Weaver, "The County as a Spatial Average in Agricultural Geography," *Geographical Review,* Vol. 46 (1956), 536–65; H. F. Gregor, "Agricultural Region and Statistical Area: A Dilemma in California Geography," *California Geographer,* Vol. 3 (1962), 27–37.

farm as simply an operating unit, when many such units are but parts of a larger ownership whose operations are centrally controlled. Sometimes even revisions of definitions are not capable of accurately representing phenomena that are rapidly changing in character, as Zelinsky noted in his attempt to use American census data for differentiating between rural farm and rural nonfarm populations.[16] On the other hand, unwanted changes have also been made over the years, making comparisons of various census reports impossible. The separation of frequency distributions into a uniform, arbitrary number of groupings often conceals the sizes of the most significant groups of large and small farms, and significant historical developments and local agricultural variations are commonly ignored because those data do not fit into a standardized statistical form. Even the time between data collections can be a problem, for the longer the span the more difficult it becomes to compute mean values and trends.

Maps and Diagrams

Maps and diagrams are both stimulants and products of field and archival research in agricultural geography. Their swelling numbers and increasing variety may thus be considered almost as good an indicator of the vigor of the field as the published articles which they usually accompany.

Land use maps are the oldest kinds of maps, and still one of the most popular among agricultural geographers. Only the aerial photograph comes closer to reality; moreover, its drawing requires no involved computations. As aerial photo coverage has increased, construction of land use maps has become easier. The additional popularity in the United States of the land use map in strip form also reflects the extensive use of the traverse in field work in this country.

The even greater popularity of thematic or special-purpose maps can be traced to the expansion of archival material. Of the two types of thematic maps, quantitative and qualitative, the first has been most preferred by American geographers, special attention being given to those using isopleths. This favoritism can be explained by the particular value of isopleth maps for analyzing data obtained through ratios. Jones first detailed this relationship when he observed that by making rough isopleth maps one would find that certain areas stood out as distinctive core areas, and further, that by comparing these maps the investigator could discover those regionalizing criteria that were more fundamental. Although now more than thirty years old, the work of Hartshorne and Dicken in regionalizing European and North American agricultural types is still probably the best illustration of what these methods can do. Another quantitative map type, the dot map, has also enjoyed wide use.

[16] M. C. Prunty, Jr., "The Census on Multiple-Units and Plantations in the South," *Professional Geographer*, Vol. 8, No. 5 (1956), 2–5; W. Zelinsky, "Changes in the Geographic Patterns of Rural Population in the United States: 1790–1960," *Geographical Review*, Vol. 52 (1962), 492–524.

Durand, in his many articles on dairy regions, has been a particularly heavy user of the technique.[17] Choropleth maps, those that present statistical data usually by political divisions, have been as popular with American geographers as isopleth maps, and even more among Europeans. Their wide currency is understandable because most statistical agencies depend on returns based on political units; moreover, in Europe administrative areas are generally smaller and archival records longer.

European geographers have also emphasized the qualitative type of thematic map more than Americans have. Again one may observe that the greater agricultural tradition in European life, as well as the longer period of geographic research, are the principal reasons for this contrast. The closely interwoven but highly variegated spatial pattern of social, historical, and economic factors influencing European agriculture has posed both opportunities and problems in cartography. This may explain why European agricultural maps commonly feature numerous pattern overlays and a variety of other symbols, sometimes bewilderingly so—at least to American eyes. Typical of such efforts is Otremba's attempt to map the "historical-social" agricultural regions of Germany by combining the distribution patterns of nine categories. European geographers have also dominated in the construction of simpler varieties of qualitative maps, although notable work has occasionally been done by Americans. In the maps of farming systems, for example, European maps are far more numerous and detailed but Whittlesey's worldwide map of agricultural regions is a landmark in American agricultural geography.[18] The same may be said for maps of rural morphology, those portraying field outlines and farmsteads. French and German geographers have been the principal workers, but works such as those appearing in two recent collections of studies of individual farms show that American interest in the rural mosaic is certainly not lacking.[19]

Differences between European and American scholars are much less marked in the matter of diagrammatic maps (cartograms) and diagrams. Each research group has contributed a goodly amount of them in relation to its total research, most of the differences in manner of representation reflecting individual diversions from the main stream rather than opposition of schools. Furthermore, all principal types of diagrams and related maps are illustrated in these contributions. Only a few of the more significant examples of each type can be mentioned here.

[17] Jones, "Ratios and Isopleth Maps"; R. Hartshorne and S. N. Dicken, "A Classification of the Agricultural Regions of Europe and North America on a Uniform Statistical Basis," *Annals of the Association of American Geographers,* Vol. 25 (1935), 99–120; A more recent illustration of the dot technique by L. Durand, Jr., is his article, "The Major Milksheds of the Northeastern Quarter of the United States," *Economic Geography,* Vol. 40 (1964), 9–33.

[18] Otremba, *Allgemeine Agrar- und Industriegeographie,* 2d ed. (Stuttgart: Franckh, 1960), p. 52; D. Whittlesey, "Major Agricultural Regions of the Earth," *Annals of the Association of American Geographers,* Vol. 26 (1936), 199–240.

[19] R. M. Highsmith, Jr., and others, *Case Studies in World Geography: Occupance and Economic Types* (Englewood Cliffs, N.J.: Prentice-Hall, 1961), pp. 1–92; R. S. Thoman and D. J. Patton (eds.), *Focus on Geographic Activity: A Collection of Original Studies* (New York: McGraw-Hill, 1964), pp. 1–59.

alue-by-area cartograms, in which shapes are distorted to conform in rea to the particular value represented, have been specialties of George nd the Woytinskys in their analyses of world agricultural relationships. more schematic version of this type of cartogram has been drawn by Iartshorne and Dicken in their semicircular version of a continent, on vhich is depicted the kind of distribution pattern of crop-livestock asociations one would generally find in areas of European culture.[20]

The diagrams include numerous kinds of graphs. The common bar graph has been used with striking effect by Jensch and Hiersemenzel to lepict the monthly progression of labor demands during the farming year, and Hartshorne and Dicken have varied the width of the bars and heir subdivisions to represent the comparative sizes and land use strucures of average farms of different areas. More recently, Blyn has used he line graph to demonstrate relationships between economic factors and rop boundaries along a cross-section from the Rocky Mountains to the ast coast, and an increasing number of geographers have begun to use yet another version of this graph, the scatter graph, in order to quantify the ovariance of two variables. More complex examples of line graphs have been created by Visher, who offers a "climograph" and by Thornthwaite, who developed a moisture-utilization diagram. The first enables one to forecast crop yields from the weather in the growing season; the second shows the yearly precipitation and how much water is required during the growing season to replace evaporation and transpiration by plants. Ehlers and a few other German geographers have shown the value of the radial graph in their use of the "cropping wheel" (*Anbaurad*), a device that allows one to show both spatial and chronological progressions of different land uses on a farm.[21]

Two-dimensional graphs, in which the areas of geometric figures are directly proportional to the values, are also represented throughout the literature of agricultural geography. The most common of these, the pie graph, is well represented by Hart and Mather's "chorographic compage map." Striking examples of the utility of the three-dimensional graph in agricultural-geographic research are the "regional diagram" used by Brown and Torbert. This graph is much like the block diagram, although the surface portrays land use patterns. The sides of the block show those subsurface aspects that are critical to agriculture, such as the position of the water table and the zonation of the soil. Related but more mathemati-

[20] P. George, *La campagne* (Paris: Presses univ. de France, 1956), pp. 41, 44; W. S. and E. S. Woytinsky, *World Population and Production* (New York: Twenieth Century Fund, 1953), pp. 540, 665; Hartshorne and Dicken, *op. cit.*, p. 120.

[21] G. Jensch, *Das ländliche Jahr in deutschen Agrarlandschaften* (Berlin: Reimer, 1957): S.–E. Hiersemenzel, *Britishe Agrarlandschaften im Rhythmus des andwirtschaftlichen Arbeitsjahres* (Berlin: Reimer, 1961); Hartshorne and Dicken, *p. cit.*, pp. 99–120; G. Blyn, "A Diagram for Demonstrating Certain Concepts in Economic Crop Geography," *Professional Geographer*, Vol. 12, No. 2 (1960), 1–3; S. S. Visher, "Weather Influences on Crop Yields," *Economic Geography*, Vol. 16 (1940), 436–43; C. W. Thornthwaite, "An Approach toward a Rational Classificaion of Climate," *Geographical Review*, Vol. 38 (1948), pp. 55–94; Ehlers, *op. cit.*, p. 176; Otremba, *Allgemeine Agrar- und Industriegeographie*, 2d ed. (Stuttgart: Franckh, 1960), p. 69.

cal versions are to be found in the "stepped statistical surfaces" used by Jenks to show the distribution of rural population density. Another version of the wedding of the map to the diagram is that of the map on which diagrams are superimposed. This cartographic device, however, has been used to only a relatively small extent by agricultural geographers because it is better suited for representing concentrated, or "point," phenomena than it is for representing the generally more spread out agricultural features. But where agriculture is intensive and highly variable, and yet occupies extremely small and scattered areas, such presentations may be ideal—as the author has found in his studies of agriculture in southern California.[22]

Agricultural maps and diagrams, besides maintaining their traditional role as accompaniments of geographic research, are becoming major objectives in themselves at a rapid rate. Increasing concern with resources, vastly improved mapping technology, and availability of funds have all helped to bring about an unprecedented increase in atlases and land use surveys. Within a ten-year period, 1954 to 1964, American, German, Russian, and English geographers, aided by fine technical staffs, have all produced agricultural atlases.[23] National atlases, such as the *Atlas of Britain and Northern Ireland* (1963) and the *Economic and Social Atlas of Greece* (1964), and government-produced agricultural publications other than atlases, such as the quinquennial *Graphic Summary* of the U.S. Census of Agriculture, also evidence the postwar cartographic advance. The land use surveys now being conducted by the various countries are also far more numerous than before World War II, the most ambitious being the World Land Use Survey, a cooperative venture by geographers of different countries for producing a land use map series at a scale of 1:1,000,000.

The Mathematical Surge

Postwar research in agricultural geography is also notable for an increasing resort to mathematical methods, partially associated, as we have already observed, with the increasing emphasis on theoretical geography. However, mathematics is not new to agricultural geography, nor are the forms used unique to the field; moreover, most of those who

[22] J. F. Hart and E. C. Mather, "The Chorographic-Compage Map," *Surveying and Mapping,* Vol. 13 (1953), 333–37; R. H. Brown, "On the Merits of Regional Diagrams in Field Reports," *Annals of the Association of American Geographers,* Vol. 25 (1935), 75–83; E. N. Torbert, "Specialized Commercial Agriculture of the Northern Santa Clara Valley," *Geographical Review,* Vol. 26 (1936), 247–63; G. F. Jenks, "Generalization in Statistical Mapping," *Annals of the Association of American Geographers,* Vol. 53 (1963), 15–26; H. F. Gregor, "A Map of Agricultural Adjustment," *Professional Geographer,* Vol. 16 (1964), 16–19.

[23] W. Van Royen, *Atlas of the World Resources,* Vol. 1: *The Agricultural Resources of the World* (Englewood Cliffs, N.J.: Prentice-Hall, 1954); E. Otremba, ed., *Atlas der deutschen Agrarlandschaften* (Wiesbaden: Steiner, 1962, 1965, 1969), Fascls. 1, 2, 3; B. Skibbe, *Agrarwirtschaftsatlas* (Gotha: Haack, 1958); M. I. Nikishov, *Atlas sel'skogo khozyaystva SSSR* [*Agricultural Atlas of the U.S.S.R.*] (Moscow: Glavnoye Upravleniye Geodezii i Kartografii SSSR, 1960); J. T. Coppock, *An Agricultural Atlas of England and Wales* (London: Faber and Faber, 1964).

ave used mathematics in this field in one way or another do not con-
.der themselves theoretical geographers in the narrow sense of the
rm. Perhaps this is why the majority of geographers using mathematical
chniques have used only descriptive mathematics, that is, descriptive
atistics that can be applied mechanically to many bodies of data, rather
ıan theory-building mathematics, whose application is not mechanical
nd whose form must be specifically designed for a particular problem in
rediction. Nevertheless, work with this elementary form of mathematics
romises to add materially to our knowledge of agricultural geography
nd to enhance the utility of other research methods as well. Bunge, for
ıstance, has demonstrated that random sampling could have radically
educed the field time that Finch spent in his intensive fractional-code
tudy. Clark greatly expanded the range of areal comparability in his
istorical study of sheep-swine ratios in Nova Scotia counties, in which
e first divides the ratio of one period by an earlier one, and then divides
his "ratio of change" in the county by the ratio of change in the province.
nd Barnes and Robinson have proposed a refinement of the isoplethic
ıethod of representing dispersed rural population distribution by sub-
tituting for the traditional people-area ratio the linear distance between
armhouses.[24]

Descriptive mathematics has also demonstrated its value in efforts
.t agricultural regionalization, particularly in the work of Weaver, Cop-
ock, Zobler, and Enyedi. Weaver has used it in his construction of crop
ombination and livestock-combination regions in the Middle West. In
is more extensive works on crop-combination regions, Weaver first com-
uted the amount of total harvested cropland occupied by each of the
rops that held as much as 1 per cent of this total in each of 1,081 coun-
ies. Next, he constructed a theoretical curve of ten crop-combination
ategories, ranging in arithmetic progression from monoculture, where
.00 per cent of the harvest is one crop, to a ten-crop combination in
vhich each of the ten crops accounts for 10 per cent of the total. Then,
gainst each of these categories Weaver measured the actual occurrence
f percentages within the individual counties, using the standard devia-
ion formula. The lowest deviation derived revealed the characteristic
rop combination, in both number and types of crops, of each of the
ounties. These designations were then plotted on maps, and boundaries
vere drawn around blocks of counties with similar crop combinations.
modified version of Weaver's procedure was later used by Coppock
o produce not only crop and livestock combinations but also combinations
f agricultural enterprises.[25]

[24] Bunge, *op. cit.*, pp. 97 f.; A. H. Clark, "The Sheep/Swine Ratio as a Guide
o a Century's Change in the Livestock Geography of Nova Scotia," *Economic Geog-
aphy*, Vol. 37 (1961), 38–55; J. A. Barnes and A. H. Robinson, "A New Method
or the Representation of Dispersed Farm Population," *Geographical Review*, Vol. 30
1940), 134–37.

[25] J. C. Weaver, "Crop-Combination Regions in the Middle West," *ibid.*, Vol.
.4 (1954), 175–200; *idem*, "Crop-Combination Regions for 1919 and 1929 in the
Aiddle West," *ibid.*, 560–72; *idem*, L. P. Hoag, and B. L. Fenton, "Livestock Units
.nd Combination Regions in the Middle West, *Economic Geography*, Vol. 32 (1956),
37–59; Coppock, "Crop, Livestock, and Enterprise Combinations."

Zobler has shown how the application of the chi-square test to field observations can meet the problem of determining whether the differences in observed values, which the investigator uses for drawing regional boundaries, are valid or simply due to chance.[26] This is done by comparing the frequency distribution of field data from each region with the frequency distribution that might be expected if no relationships existed between the regions and the test information. If the difference exceeds a certain value, the hypothesis that no relationship obtains between the regions and the test data is rejected, and the remaining alternative, the regional hypothesis, is accepted; if the difference is less, the regional hypothesis is rejected, for the regional differences noted in the field reports, vivid as they are, may have been the result of chance sampling.

To discover where the agricultural production is concentrated within a region, Eastern Europe, Enyedi devised a coefficient that enabled him to compare the share that each Eastern European country had of the total production and territorial area. If the coefficient was greater than 1, the country had a larger share of production than the regional average. He then set up indices with which he attempted to discover whether these regional differences were declining or increasing.[27]

A more complex coefficient, one traditional in mathematics, is produced by regression analysis and related correlation studies. These techniques come into play when there is a need to compare sets of data in terms of the extent to which a change in one set is or is not reflected by a change in the other. The search for coefficients of correlation in agricultural geography, as old as the field itself, has traditionally concentrated on associations between crops and climatic factors. More recently, there has been a strong urge to calculate these coefficients for other types of associations, such as between cash-grain farming and landforms, farm population density and farm size, and farm population density and precipitation. The study of this last association, by Robinson and Bryson, is especially interesting in that it demonstrates the possibility of mapping the amount of spatial covariation of two or more distributions.[28]

The ultimate in sophisticated methods of use to agricultural geography is the mathematical model. This is the term frequently applied to a hypothesis that is stated in such a way that the effect of changes in any of its independent variables can be assessed. In their more highly developed forms, models are capable of expressing relationships which are so complex that they would be practically impossible to contemplate if they were stated in nonmathematical terms. These more detailed models

[26] L. Zobler, "Statistical Testing of Regional Boundaries," *Annals of the Association of American Geographers,* Vol. 47 (1957), 83–95.

[27] G. Enyedi, "The Changing Face of Agriculture in Eastern Europe," *Geographical Review,* Vol. 57 (1967), 358–72.

[28] A. H. Robinson and R. A. Bryson, "A Method for Describing Quantitatively the Correspondence of Geographical Distributions," *Annals of the Association of American Geographers,* Vol. 47 (1957), 379–91; A. H. Robinson, J. G. Lindberg, and L. W. Brinkman, "A Correlation and Regression Analysis Applied to Rural Farm Population Densities in the Great Plains," *ibid.,* Vol. 51 (1961), 211–21.

of agricultural activity have been used much more by agricultural economists and for a longer time, but geographers are beginning to employ them increasingly. Further, they are using such models in an extremely wide variety of ways, ranging from providing a proof for the theorem that for every spatial location there is some jointly optimum intensity of land use to programming the most efficient use of space for food production.[29] There is also increasing interest in models dealing with behavioral, as well as economic, problems. A monograph on the "diffusion of the decision to irrigate" among Great Plains farmers and an article on the "decision process in spatial context" among farmers of Middle Sweden are two of the latest examples of this trend.[30] Still another path for more intensive model research appears to lie in the construction of stochastic models, those that deal with change, in contrast to the traditional static equilibrium models.[31]

As methods like mathematical model-building and regression analysis provide the means by which the geographer can study a larger number of variables, they also make it necessary for him to have a way of comparing, calculating, and sorting this information rapidly. High-speed data-processing equipment is therefore becoming another valuable research aid, although its expense and the need for expert programmers still hinder its wider use. As yet, only Coppock has given a specific and practical account of the value and problems of computerized research in agricultural geography.[32]

Will mathematics eventually supplant the map as the most characteristic research tool of geography? Many mathematically inclined geographers apparently believe so. Yet, work by such men as Weaver and Coppock has shown that mathematically based research in agricultural geography can itself lead to a large and often unusual series of maps. At the same time, most of the research of this type has so far been confined to English-speaking geographers; in France and Germany, the more historical and cultural orientations seem to have made mathematical methods less attractive. A surer answer to the question, then, will have to await future trends. In the meantime, mathematics will undoubtedly reinforce its claim to an already respectable position among the major research methods of geography.

[29] W. L. Garrison and D. F. Marble, "The Spatial Structure of Agricultural Activities," *ibid.*, Vol. 47 (1957), 137–44; and L. Zobler, "A New Areal Measure of Food Production Efficiency," *Geographical Review*, Vol. 51 (1961), 549–69.

[30] L. W. Bowden, *Diffusion of the Decision to Irrigate* (Chicago: Univ. of Chicago, 1965); J. Wolpert, "The Decision Process in Spatial Context," *Annals of the Association of American Geographers*, Vol. 54 (1964), 537–58.

[31] D. W. Harvey, "Theoretical Concepts and the Analysis of Agricultural Land-Use Patterns in Geography," *ibid.*, Vol. 56 (1966), 374.

[32] J. T. Coppock, "Electronic Data Processing in Geographical Research," *Professional Geographer*, Vol. 14, No. 4 (1962), 1–4; *idem* and J. R. Johnson, "Measurement in Human Geography," *Economic Geography*, Vol. 38 (1962), 130–37.

PART 2 the study of landscape

CHAPTER THREE

The role of the environment

The Significance of the Environment

Certainly no other branch of human geography concerns itself so directly with the physical environment as agricultural geography. From the first, workers in the field were attracted to the problem of explaining how variations in the environment affected the agricultural landscape. However, as in the methodological consideration of the context in which areal variation of agriculture is to be studied, approaches to the study of the relations between agriculture and environment have been varied and changing.

The oldest approach has emphasized more the environmental side of these relationships, especially the effects of climate on the growth of crops. Research on this topic has progressed so far that several subfields of interest can be recognized. Climatic variations and their effect on yields have received the most attention. True to their orientations, agriculturists, e.g., Bennett and Timoshenko, have studied this type of relationship more in terms of a single crop and its world pattern, whereas geographers have more often considered one or several crops of a particular region. Thus, Rose has analyzed the relationship between corn yield and climate in the Corn Belt, Weaver has investigated the relationship between barley and climate in South Dakota, Malmström has studied the connections between potatoes and the Arctic Front in Iceland, and Barton and Hore have studied rainfall and rice in Thailand and West Bengal, respectively. Visher, however, has expanded his investigation of relations to four crops in Indiana, and Maunder has studied the total agricultural production of New Zealand. Mathematical methods were first used in agricultural geography in studies of this type, and those of Maunder represent some of the most advanced. He used a multiple regression model to assess the association between twenty-three agricultural factors and fifteen aspects of the climate. That such complex methods are not always needed for correlation studies, however, is well shown by the work of Hewes in determining the causes of wheat failure in the Great Plains.[1]

Only slightly less abundant in the literature on significant relationships between climate and crops have been the reports on the various limits imposed on the growth of crops by temperature and moisture. A related effort has been the mapping of the areal progression of seasonal biological phenomena, such as the onset of flowering and ripening. The numerous phenological maps of the Department of Agriculture in the United States and those of Schnelle in Europe show that this type of research continues in full swing.[2] In turn, the expansion of these two kinds of information has provided a basis for increased efforts to outline the areas which are ideal for various crops.

From agricultural science has come a major effort to set up a world climatic classification based on the suitability of specified crops for various regions. Bennett has correlated the boundary values described by geographers with the requirements of the major food crops, to arrive at a map of "foodcrop climates." Papadakis has produced an unusually detailed map of this type, "Climates of the World and Their Agricultural Potentialities," containing seventy climatic subdivisions. Much the same work may also be observed in the numerous publications by Nuttonson in his regional "climatic analogs." Meanwhile, more local but also more intensive studies go on, as agriculturists seek to delimit the optimum areas for valuable crops. One of the best examples of this work has been the use of regression analysis in determining the "natural location" of corn areas in West Germany, by Liesegang and Schall.[3]

[1] M. K. Bennett, "Trends of Yield in Major Wheat Regions since 1885," *Food Research Institute Wheat Studies,* Vol. 14, Nos. 3 (1937) and 6 (1938); V. P. Timoshenko, *World Wheat Production: Its Regional Fluctuations and Interregional Correlations* (Stanford, Calif.: Stanford Univ., 1944); J. K. Rose, "Corn Yield and Climate in the Corn Belt," *Geographical Review,* Vol. 26 (1936), 88–101; J. C. Weaver, "Climatic Relations of American Barley Production," *ibid.,* Vol. 33 (1943), 596–88; V. H. Malmström, "Influence of the Arctic Front on the Climate and Crops of Iceland," *Annals of the Association of American Geographers,* Vol. 50 (1960), 117–22; T. F. Barton, "Rainfall and Rice in Thailand," *Journal of Geography,* Vol. 62 (1963), 414–18; P. N. Hore, "Rainfall, Rice Yields and Irrigation Needs in West Bengal," *Geography,* Vol. 49 (1964), 114–21; S. S. Visher, "Weather Influences on Crop Yields," *Economic Geography,* Vol. 16 (1940), 436–43; W. J. Maunder, "Climatic Variations and Agricultural Production in New Zealand," *New Zealand Geographer,* Vol. 22 (1966), 55–69; L. Hewes, "Causes of Wheat Failure in the Dry Farming Region, Central Great Plains: 1939–1957," *Economic Geography,* Vol. 41 (1965), 313–30.

[2] F. Schnelle, "Phänologische Europakarten: Beginn der Apfel-Blüte und Beginn der Winterweizen-Ernte," *Erde,* Vol. 97 (1966), 138–44.

[3] M. K. Bennett, "A World Map of Foodcrop Climates," *Food Research Institute Studies,* Vol. 1 (1960), 285–95; J. Papadakis, *Climates of the World and Their Agricultural Potentialities* (Buenos Aires: Privately printed, 1966); M. Y. Nuttonson, "Crop and Weather," *Landscape,* Vol. 12 (1962), 9–11; F. Liesegang and S. S. Schall, "Der natürliche Standort für den Anbau von Körner- und Silomais in der Bundesrepublik Deutschland," *Berichte über Landwirtschaft,* Vol. 44 (1966), 568–605.

The regional delimitation of climates suitable for crops has also continued to interest a certain number of geographers. Sapozhnikova and Shashko reflect the enthusiasm of Soviet agricultural geographers for this topic in their mapping of the "agroclimatic belts" of the U.S.S.R., and Otremba has discussed in detail the values of "plant-capability" areas for geographic research. But nothing as yet in the geographic literature has surpassed the originality and promise of Thornthwaite's method of classifying climates. It is based on the regional variation of "potential evapotranspiration," the combined evaporation from the soil surface and transpiration from plants. Thornthwaite and his co-workers discovered that crop growth was closely related to this process, and that therefore the annual potential evapotranspiration provided an index of the growth potential of an area. Thus, given records of the temperature and precipitation of an area and the water requirements of the individual plants, it became possible to select the crops most suited to the area or to calculate the extra water needed if irrigation were necessary.[4] Thornthwaite's scheme has not escaped criticism. The most serious objection has been that air temperature, which Thornthwaite used to express potential evaporation, plays a much lesser role than radiation. Some climatologists also claim that his climatic classification is excessive and cumbersome and fails to bring out many important climatic features.[5] As with all research successes, therefore, the path now lies open for even further advances.

Agriculturists, and geographers even more, have also been concerned with the discrepancies between the ecological boundaries of crops and those delimiting the actual crop margins. Mead, for example, has mapped the disconformities between economic and "absolute" crop boundaries in Finland, noting large areas still open to wheat expansion. Ackerman has observed that commercial citrus acreage straddles its limit of safe growth in response to middle-latitude markets, thus becoming increasingly uneconomic. That such locations make agriculture more costly is debated by some who, like Huntington, maintain that crop areas on cooler margins also reflect certain environmental advantages, such as greater variability of temperature, longer periods of sunlight, and greater freedom from blights and insects. Yet most of the arguments over the comparative importance of physical and economic limits of crop distribution have stopped short of considering the location of still another powerful barrier to the spread of crops, political and ethnic influences. One of the few attempts to show how these different types of boundaries actually compare with each other in their relation to the distribution of a crop

[4] S. A. Sapozhnikova and S. I. Shashko, "Agroclimatic Conditions of the Distribution and Specialization of Agriculture," *Soviet Geography: Review and Translation,* Vol. 1, No. 9 (1960), 20–35; E. Otremba, "Agrarische Wirtschaftsräume, ihr Wesen und ihre Abgrenzung," *Akademie für Raumforschung und Landesplanung,* Vol. 20 (1962), 5–20; C. W. Thornthwaite, "An Approach toward a Rational Classification of Climate," *Geographical Review,* Vol. 38 (1948), 55–94.

[5] J. H. Chang, "An Evaluation of the 1948 Thornthwaite Classification," *Annals of the Association of American Geographers,* Vol. 49 (1959), 24–30.

has been made by Deasy, in his delimitation of areas favorable for commercial coconut production.[6]

Considerably less has been written by geographers on the effects of climate on livestock. In American agricultural geography, one must go back to the writings of the 1920's, such as Davidson's.[7] This kind of work now occurs almost wholly within zoology, animal geography, and animal genetics. The contributions from these fields, however, as well as the references still made by geographers in connection with other topics, have provided enough material for occasional discussions.[8]

Quite different is the picture if one considers the work of geographers in analyzing the influence of terrain in agriculture. Here a long interest is on record, although geographers are now much less inclined to accord terrain a weighty role in land use variations. Every major terrain type of significance to agriculture has been treated, ranging from individual landforms to extensive terrain complexes. Plains, quite naturally, have received abundant attention. Their intimate relationship with agricultural activity prompted Hidore to make what is perhaps the first attempt to ascertain statistically the extent of association of flat land in the Midwest with cash grain farming, the results of which are as interesting in their deviations as in their concordances. Particular features of plains, such as natural levees and deltas, have also been intensively studied, especially in Europe and Southeastern Asia, where land use patterns are so intricate and finely adapted to the surface configuration. Dobby's detailed correlation of settlement and agricultural patterns with the "ridge and hollow" terrain left by meandering streams on the Kelantan Delta in Malaya and the description and large-scale mapping by Burger of field patterns on a portion of the Rhine lowland, are two especially impressive examples of the work done on these continents. The use of naturally terraced land has also been given close scrutiny. Among American writers, Trewartha has particularly distinguished himself by his studies of the historical development of land use on the Prairie du Chien terrace in Wisconsin, and by his detailed contrasts between the agriculture of the flood plains and the diluvial terraces in Japan.[9]

The agricultural role of plains modified by glaciation has also been studied on an intensive scale. Some geographers have viewed the glacial landform complex in its entirety, as in Millington's article on horticulture

[6] W. R. Mead, "Agriculture in Finland," *Economic Geography,* Vol. 15 (1939), 219; E. A. Ackerman, "Influences of Climate on the Cultivation of Citrus Fruits," *Geographical Review,* Vol. 28 (1938), 289–302; E. Huntington, *Principles of Human Geography* (New York: Wiley, 1940), pp. 233–40; G. F. Deasy, "Location Factors in the Commercial Coconut Industry," *Economic Geography,* Vol. 17 (1941), 130–40.

[7] F. A. Davidson, "Relation of Taurine Cattle to Climate," *ibid.,* Vol. 3 (1927), 466–85.

[8] E.g., P. Veyret, *Géographie de l'élevage* (Paris: Gallimard, 1951), pp. 29–34.

[9] J. J. Hidore, "Relationship Between Cash Grain Farming and Landforms," *Economic Geography,* Vol. 39 (1963), 84–89; E. H. G. Dobby, "The Kelantan Delta," *Geographical Review,* Vol. 41 (1952), 226–55; G. Burger, "Weisweil, Kreis Emmendingen, 1958," in *Atlas der deutschen Agrarlandschaft,* ed. E. Otremba (Wiesbaden: Steiner, 1962), 1st Lfg., Teil V, Blatt 1-d; G. T. Trewartha, "The Prairie du Chien Terrace: Geography of a Confluence Site," *Annals of the Association of American Geographers,* Vol. 22 (1932), 119–58; *idem, Japan* (Madison: Univ. of Wisconsin, 1945), pp. 22–26, 471–77.

in central Massachusetts and Lee's and Reed's writings on agricultural regions in Ontario. Others, such as Durand and Dahlberg, have been more interested in the agricultural significance of only certain glacial landforms, such as drumlins or lacustrine plains and Pleistocene beaches.[10]

Mountain-valley complexes and their often intricate land use patterns have attracted the attention of far more investigators than their economic importance would indicate. French geographers, who have done numerous regional studies in the Alps and Jura, have been the most active among this large category of researchers. As in the investigations of land use on glaciated plains, the researchers have dealt with both the over-all terrain and individual landforms. This is true of North American geographers, too, even though the proportion of their research in agricultural geography devoted to mountain areas has been considerably less than that of their European colleagues. Taking the more comprehensive view, Kerr has considered the implications of the intricate physical patterns of British Columbia for over-all agriculture, and Beard has similarly considered a part of that section of the California Coast Ranges which fronts on the ocean. Krueger and Weir, on the other hand, have restricted their studies of the significance of the mountain-valley complex to one agricultural category—Krueger to orchards, and Weir to livestock. Still others have confined their work to deciphering vertical zonation patterns of land use in mountainous regions, although Peattie has rejected these patterns as too coarse a generalization. Of the studies dealing with smaller physiographic units, Hall's on the farming system in the Yamato Basin in Japan and Trewartha's on the apple economy of the Iwaki Basin in the same country remain classics. The same may be said of the descriptions by Colby and Torbert of land use zonation on alluvial fans in California.[11]

[10] B. R. Millington, "Glacial Topography and Agriculture in Central Massachusetts," *Economic Geography,* Vol. 6 (1930), 408–15; C. Lee, "Land Utilization in the Middle Grand River Valley of Western Ontario," *ibid.,* Vol. 20 (1944), 130–51; L. G. Reeds, "Land Utilization in Central Ontario," *ibid.,* Vol. 22 (1946), 289–306; L. Durand, Jr., "The Cheese Region of Southeastern Wisconsin," *ibid.,* Vol. 15 (1939), 284–92; R. E. Dahlberg, "The Concord Grape Industry of the Chautauqua-Erie Area," *ibid.,* Vol. 37 (1961), 150–69.

[11] D. Kerr, "The Physical Basis of Agriculture in British Columbia," *ibid.,* Vol. 28 (1952), 229–39; C. N. Beard, "Land Forms and Land Use East of Monterey Bay," *ibid.,* Vol. 24 (1948), 286–95; R. R. Krueger, "The Physical Basis of the Orchard Industry of British Columbia," *Geographical Bulletin,* Vol. 20 (1963), 5–38; T. R. Weir, *Ranching in the Southern Interior of British Columbia,* Memoir 4, Geographical Branch, Mines and Technical Surveys, 2d ed. (Ottawa: Queen's Printer, 1964); Studies of vertical zonation are H. J. Critchfield, "Land Use Levels in Boundary County, Idaho," *Economic Geography,* Vol. 24 (1948), 201–208; and H. J. Cox, "Weather Conditions and Thermal Belts in the North Carolina Mountan Region and Their Relation to Fruit Growing," *Annals of the Association of American Geographers,* Vol. 10 (1920), 57–68. Cf. R. Peattie, "Height Limits of Mountain Economies," *Geographical Review,* Vol. 21 (1931), 415–28.

R. B. Hall, "The Yamato Basin, Japan," *Annals of the Association of American Geographers,* Vol. 22 (1932), 252; G. T. Trewartha, "The Iwaki Basin: Reconnaisance Field Study of a Specialized Apple District in Northern Honshu," *ibid.,* Vol. 20 (1930), 196–223; C. C. Colby, "Piedmont Plain Agriculture in Southern California," in *An Introduction to Economic Geography,* ed. W. D. Jones and D. S. Whittlesey (Chicago: Univ. of Chicago, 1925), Vol. 1, pp. 218–21; E. N. Torbert, "Specialized Commercial Agriculture of the Northern Santa Clara Valley," *Geographical Review,* Vol. 26 (1936), 247–63.

As works on terrain-agriculture relationships have increased, so also have efforts at quantifying these relationships, expanding study areas, and developing typologies. To the work of Hidore on the correlation of flat lands and cash grain farming may be added the earlier attempt by Garnett to correlate mathematically the interrelationships of isolation, topography, and human life [*sic*] in mountain lands. MacGregor has asked for more statistical precision in reporting on correlations between slope and land use in Britain, and has suggested several specific categories of slope as being critical. Marschner has shown how a rich source of aerial photographs can enable a single researcher to study sample land use patterns over an extensive area, in this case, the United States. Burton has proposed a classification of floodplains based on "occupance types," each of which is defined by a particular combination of flood-plain width, slope of the adjoining land, flood frequency, flood seasonality, land use intensity, and location of farm structures relative to the flood plain. Of the many possible combinations, twenty-seven are suggested as being most likely to occur.[12]

Research on the influence of soil on agricultural patterns has been less intensive than on climate and terrain, but enough has been written to display a number of different interests and trends. As in the search for areas of ideal crop growth based on climate, agriculturists and geographers have been concerned with finding the most suitable soil areas, or "edaphic optima," for individual crops. A corollary of many of these studies is the comparison of optimum edaphic and climatic areas in order to find the most productive soil for the crop, as Weaver does for barley. Numerous data are now also available to show us that animals, as well as crops, have their optimum soil area. Their growth and the amount and quality of their products depend strictly on the quality of their forage, which, except for carbon dioxide and the nitrogen used by legumes, draws all its materials from the earth. From comparisons of the distributions of soil minerals and livestock we learn that soil composition must now be considered on a par with other physical and cultural determinants of livestock location, as Veyret's work in France illustrates so well.[13]

From the study of correlations between soils and crops and animals, it has been but a step to a consideration of the most critical association of all, that of soil quality and the condition of man. A growing number of investigators have now challenged the older theories of racial differences, claiming that differences in physique are a result of regional adjustments to certain soil deficiencies which exert their influence through crop and animal products. Indeed, Castro has gone so far as to assert that many of the afflictions that have scourged man and which traditionally have

[12] A. Garnett, "Topography and Settlement in the Alps," *ibid.*, Vol. 25 (1935), 601–14; D. R. MacGregor, "Some Observations on the Geographical Significance of Slopes," *Geography*, Vol. 42 (1957), 167–73; F. J. Marschner, *Land Use and Its Patterns in the United States* (Washington, D.C.: U.S. Govt. Printing Office, 1959); I. Burton, *Types of Agricultural Occupance of Flood Plains in the United States* (Chicago: Univ. of Chicago, Department of Geography, 1962).

[13] J. C. Weaver, *American Barley Production* (Minneapolis: Burgess, 1950); Veyret, *op. cit.*, p. 25.

been ascribed to climate or considered racial characteristics are actually due to these soil deficiencies. Most geographers, however, have not been willing to give this priority to soil, preferring rather to view it as a secondary agent in the agricultural process. Thus, Maxwell has emphasized how even in the Ontario Shield, where the availability of soil might be expected to be the simple and ultimate determinant of the cropping pattern, farming technology and concepts of living encouraged farmers to select only certain soils; as technology and living standards changed, soils once considered the best were rejected for another kind that had been previously considered too difficult to work.[14]

Adjustments in Farming Operations

The shift by geographers toward a greater emphasis of the human role in their study of the relationships between agriculture and environment is even more evident, however, in the increasing attention being given to the ways men have adjusted farming operations to the variations and vagaries of the environment. In this area there are more geographers, in relation to agriculturists, than there are in the more physically oriented studies. Here, too, the agricultural economist and rural sociologist are more active than the agronomist, geneticist, and other specialists in the physical sciences. But this emphasis is not without precedent, nor is there a sharp line between the work of the two groups. As will become increasingly clear in our review of research themes in agricultural geography, investigators, particularly geographers, have not uncommonly been concerned with several aspects of the agriculture of an area, the relative priorities of which, in any one work, are not always easy to discover.

The good side of this multiple view, however, is that the significance of any one aspect of agriculture is greatly enhanced by the consideration of as many related aspects as possible. A case in point is the research done on the way the farmer adjusts his work cycle to the climate. By itself, this relationship is basic enough, but how it is modified by elements other than climate is also critical. Thus Torbert shows how the yearly schedule of farming operations within a small California valley can vary, depending on whether the farm is on the valley flat, alluvial fan slope, or hill and mountain slope. Even where differences in terrain are not extreme, differences in crop emphasis can produce a variety of farming cycles within a restricted area, as shown in Schroeder's descriptions of cotton, vegetable, and citrus farms in the Rio Grande delta. The normal influences of seasonal climatic rhythm can also be directly contradicted by the distance from markets, as Stephan has shown happens in the Florida Everglades. Here the hottest and wettest time of the year for

[14] J. de Castro, *Géopolitique de la faim,* 2d ed. (Paris: Editions ouvrières, 1962), pp. 70–71; J. W. Maxwell, "Notes on Land Use and Landscape Evaluation in a Fringe Area of the Canadian Shield," *Geographical Bulletin,* Vol. 8 (1966), 134–50.

vegetable raising is also the period of least activity because the northern markets are monopolized by local northern producers at that time.[15]

The operational complexities of the farming year have been a problem as well as an attraction to all who have studied them and wished to present them clearly. Trewartha has given a good idea of how difficult this situation can be, in a study of Japanese agriculture in which he observes how multiple cropping and intercropping combine in a bewildering variety of ways. His maps of "annual frequency of cultivation," obtained by dividing the total area of cultivated land into the total area of all crops grown during a given year, are attempts to summarize this high level of intensity in an entire country. The closest approach to reality, of course, is the mapping of the crop landscapes at various stages of the cropping calendar. By mapping only about forty acres, but at four different times, Thompson was able to determine the changes in the percentage of land allocated to each crop. Other researchers have felt that more meaningful measures of activity in the farming year are the type and amount of work performed by the farmer. Jensch has applied the criterion of "man-work hours" (MAS–*Menscharbeitstünden*) to the main farming operations on several West German farms, and in the process has constructed graphs showing the seasonal progression of these activities and whether they are carried on indoors or outdoors. Hiersemenzel, a student of Jensch, has done the same with several English farms. In addition, she has mapped the monthly distribution of the various farming activities over the fields of individual farms. Thus a way has been provided for greater regional comparison of farming years, and in good part on a uniform statistical basis. These have not been the only recent achievements in methods of making regional comparisons of farming years, certainly. All depends on the particular purpose of the project, as Curry demonstrated in his correlation of "seasonal programming" of New Zealand livestock farms with the evapotranspiration and dry-matter-production probability regime of various regions.[16]

It is to Spencer and Deffontaines, however, that credit must probably be given for focusing attention on what is perhaps currently the most intriguing question about the farming year: Can there be a climatically determined round of agricultural activities where there is little seasonality? It has been customary to assume that planting is nonseasonal in the tropics, but Spencer has described many crop calendars that are followed by cultivators in equatorial latitudes. This raises the further question of whether these schedules are cultural in timing or actually adjustments to microclimatic controls. Up to now, the lack of precise

[15] Torbert, *op. cit.*, p. 257; K. Schroeder, *Agrarlandschaftsstudien in südlichsten Texas* (Frankfurt am Main: Kramer, 1962); L. L. Stephan, "Vegetable Production in the Northern Everglades," *Economic Geography*, Vol. 20 (1944), 1.

[16] Trewartha, *Japan*, pp. 204–206; J. H. Thompson, "Urban Agriculture in Southern Japan," *Economic Geography*, Vol. 33 (1957), 224–37; G. Jensch, *Das ländliche Jahr in deutschen Agrarlandschaften* (Berlin: Reimer, 1957); S. E. Hiersemenzel, *Britische Agrarlandschaften im Rhythmus des landwirtschaftlichen Arbeitsjahres* (Berlin: Freien Univ. Berlin, 1961); L. Curry, "Regional Programming in the Seasonal Programming of Livestock Farms in New Zealand," *Economic Geography*, Vol. 39 (1963), 95–118.

information about the history of regional varieties of crops or about the maturity rates of specific varieties of crops has prevented us from arriving at a definitive answer. But Spencer does point out that climatically determined seasonality is indeed spreading in the tropics as shifting cultivators discard tubers, roots, rhizomes, and rootstock crops in favor of seed-planted grains, pulses, grams, seed legumes, and seed-sown vegetables. To suggest that an active farming calendar can also be found where climatic conditions are almost the exact reverse of tropical poses an even greater challenge to common sense, but Deffontaines has come close in his detailed description of the effects of an unusually long nonproductive season on farming activity in Quebec. Activities other than those connected with crops are by no means lacking in the fallow period, and the shortness of the growing season itself promotes an almost frenetic rate of change from one farming activity to another. The overlapping of seeding and harvesting, for instance, bears a striking resemblance to the dovetailing of operations characteristic of Oriental farmers.[17]

A corollary of regional variations in the farming year is the periodic shifting of production centers. Geographers have been quick to note that the great distances between the large consuming centers of Northwestern Europe and Northeastern United States and the areas with longer growing seasons but smaller populations have promoted an accordion-like movement of crop production in agreement with the progression of seasons. Much has happened since World War II, however, to modify these patterns, particularly in the United States, and many agricultural economists have referred to these developments. Few, however, have tried to summarize them in the light of their effects on national distribution patterns; the best review of their geographic consequences so far is probably Biehl's. Analyzing the rich U.S. Agricultural Census data, he found that the latitudinal staggering of truck-crop production characteristic of the Atlantic Seaboard and the Mississippi Valley has been noticeably modified by the increased competition of more-distant California with its Southeastern subtropical counterpart. Its development is a result of the deep-freezing process. Furthermore, this new technique of preservation has strengthened the heretofore deteriorating competitive position of the producing areas around the large markets of the Northeast and Midwest, thus further reducing the dependence of these areas on the warmer Southeast.[18]

The question of what kind of seasonal shifts agricultural production makes in areas where long growing seasons and large markets coincide would seem to offer little prospect for research. Yet the little that has been done seems to show that the biggest problem is not so much the lack of such periodic movements as it is their complexity. In an investigation of the movements of centers of supply of fresh fruits

[17] J. E. Spencer, *Shifting Cultivation in Southeastern Asia* (Berkeley: Univ. of California, 1966), pp. 40–47, 122–25; P. Deffontaines, *L'homme et l'hiver au Canada* (Paris: Gallimard, 1957).

[18] M. Biehl, *Der Obst-, Gemüse- und Gartenbau im Nordosten der Vereinigten Staaten von Amerika unter der Konkurrenz subtropischer Landesteile* (Kiel: Institut für Weltwirtschaft Univ. Kiel; 1958), pp. 74–119 and 149–56.

and produce for the San Francisco Bay area, for example, the writer observed that peak tomato production in the spring moves both northward (Imperial Valley to the Bay area) and coastward (Imperial Valley to the San Diego area). At the same time, the first oranges of the season come from the north, the Sacramento Valley; later, they come from the south, southern coastal California. One might expect such divergences to fade as he enters the tropics, but some, including Spencer, have noted that tropical trees that produce fruits of economic value have distinctly seasonal, though irregular, patterns of ripening, and large volumes of fruit at certain times are not to be counted on. Moreover, this irregularity has also contributed to the better-defined seasonal shifts of production peaks in higher latitudes by encouraging retailers in such places as Singapore, Malaya, and the Philippines to purchase a large variety of fruits, and even certain vegetables, from the cooler areas.[19]

Less has been written by geographers about the shifts of agricultural populations that have accompanied the periodic movements of peak production. One of the problems undoubtedly has been lack of statistics. Pelzer is one of the few who has tried, despite this deficiency, to ascertain the types and strengths of periodic flows of rural population, in this case of Southeast Asian plantation workers. Unfortunately, the political and economic upheavals during and since World War II have outdated much of the material of this kind, although United States government economists have kept up with the seasonal farm labor movements that have steadily increased since the war. We now know that three major north-south flows characterize the seasonal movements of American farm laborers. The largest of these begins in southern Texas in the winter and fans out over the Middle West and Great Plains and into the Pacific Southwest. Another originates in southern California and gradually extends northward into the Pacific Northwest. A third starts in Florida and reaches New England before it begins its southward ebb after the summer. Many individual migrants, however, move in a much less orderly way than the total, many going only part of the way or interrupting their movement for a vacation.[20]

Quite a different picture is presented in the research on the seasonal movements of graziers and their livestock. The writings have been voluminous (predominantly French and German), yet the defining of these practices has continued to generate considerable controversy. Not until 1937 was a fairly clear differentiation made, by Merner, between the three principally recognized types of shifting animal husbandry: transhumance, summer mountain pasture economy (*Almwirtschaft*), and

[19] H. F. Gregor, "The Local-Supply Agriculture of Southern California," *Annals of the Association of American Geographers,* Vol. 47 (1957), 268–69; J. E. Spencer, "Seasonality in the Tropics: The Supply of Fruit to Singapore," *Geographical Review,* Vol. 49 (1959), 475.

[20] K. J. Pelzer, *Die Arbeiterwanderungen in Südostasien* (Hamburg: de Gruyter, 1935). The work of W. H. Metzler and F. O. Sargent, *Migratory Farmworkers in the Midcontinent Streams* (Washington, D.C.: U.S. Govt. Printing Office, 1960), is typical of the government research.

nomadism. In 1960, Beuermann felt compelled to attack the problem again; he stressed the lack of a forced movement due to dryness in the lowlands as the major differentiator of the *Almwirtschaft* from transhumance. Nomadism, Beuermann reaffirmed, contrasts with transhumance in its movement of the entire family with the animals and its complete lack of cropping operations and sedentary settlements.[21] But the problems posed by innumerable borderline cases remain, such as nomads who assign the care of their herds or flocks to shepherds or transhumants who decide to move with their entire families.

Most researchers have preferred to concentrate more on the characteristics than on the boundaries of each of the three major forms of shifting animal husbandry, admitting that the ultimate delineations are at best a long way off. The biggest part of these studies has dealt with transhumance, the most common and widespread of the three forms. This large volume of contributions has been plumbed by Hofmeister in a major reevaluation of the definition of transhumance. After an exhaustive survey of previous definitions, he came to the conclusion that one or more objections may be raised against many of them. They are that: (1) transhumance has often been designated as a seasonal movement of just sheep herds; (2) summer and mountain pastures are frequently considered as one and the same; and (3) all types of shifting animal husbandry exhibiting stall feeding are included within the *Almwirtschaft* category. Hofmeister then proposed six types of transhumance: (1) *normal,* where movement is from a homestead in the lowlands to mountain pastures at the approach of the hot, dry seasons, and in reverse at the approach of winter; (2) *inverted,* where movement is from a highland farmstead to the lowlands in the winter and back again in the summer; (3) *hibernal* or *tropical,* in which the movements are always up slope as the dry winter season approaches and down slope as the wet summer season comes on, but which may be further classified as *normal hibernal* or *inverted hibernal,* depending on the location of the farmstead relative to these movements; (4) *petite,* in which animals are moved up slope during the summer and down slope during the winter, but are still kept up in the mountain areas; (5) *partial* or *mixed,* where animals are stall fed for a season because of economic rather than climatic, considerations; and (6) *complex,* where more than two seasonal pastures are used and where the farmstead is situated near one of the intermediate pastures.

Hofmeister summarized these types in a comprehensive definition of transhumance, describing it as a system of semi-nomadic livestock farming with migration or transport of the stock normally between two, occasionally more, only seasonably usable pasture grounds the locations of which differ in altitude, climate, and vegetation. Winter stall feeding may occur, but only because of cropping, not climatic, requirements. The permanent farmstead, nearly always located at one of the seasonal pasture

[21] P. G. Merner, "Das Nomadentum im nordwestlichen Afrika," *Berliner geographische Arbeiten,* Vol. 11 (1937); A. Beuermann, "Formen der Fernweidewirtschaft (Transhumance–Almwirtschaft–Nomadismus)," *Verhandlungen des deutschen Geographentages,* Vol. 32 (1960), 277–90.

grounds, and the pasture grounds used during the other season are not organically linked.[22]

Another work by Hofmeister points up the regional gaps in transhumance research just as effectively as his more methodological article reflects the over-all wealth of contributions to this research topic. This is his doctoral dissertation on transhumance in the western United States, completed in 1958. It is easily the most detailed work by a geographer on transhumance in this country, a topic which has received no attention from geographers except for only an occasional contribution, such as White's "Transhumance in the Sheep Industry of the Salt Lake Region," published in 1926. Hofmeister's research shows that source material is not lacking, however, as his many citations of federal and state publications alone indicate. The equally striking paucity of geographic writings on transhumance in areas such as Latin America and Australia are more explicable because of the very slow development of the livestock practice itself. Yet recent articles like those on transhumance in Tasmania, by Scott, and on Latin America, by Deffontaines, show that at least it is not being completely overlooked in the less typical regions.[23]

If American geographers have insufficiently investigated transhumance in their country, they have more than done their share in the intensive study of another subject: the increasingly closer adjustment of agriculture to areas with the best physical environment. Baker concerned himself with this process as early as 1921, when he described in detail many of the ways land use patterns in the United States were being increasingly patterned after the variations in terrain, soils, temperature, and moisture. Farming was becoming more extensive on the poorer lands, but more intensive on the better lands. Individual crops were being restricted to the areas particularly suitable for them. A closer correspondence between crop boundaries and those of optimal growth was being sought. All this was due to the fact the American population was now increasing faster than the supply of land, according to Baker, and thus, as land became scarcer and more valuable, greater and greater care had to be exercised in using each kind of land for the purpose most favored by the physical conditions. This growing disparity and the reaction to it was certainly not peculiar to the United States. What was impressive, however, was that all this was now occurring in an area where all the conditions favoring closer adjustment to physical differences were at their optimum: many physical environments in a subcontinental framework; a well-developed transportation system; a dominantly commercial

[22] B. Hofmeister, "Wesen und Erscheinungsforem der Transhumance," *Erdkunde,* Vol. 15 (1961), 121–35. I. M. Matley, in his article "Transhumance in Bosnia and Herzegovina" [*Geographical Review,* Vol. 58 (1968), 230–36], challenged Beuermann's definition, but makes no mention of Hofmeister's detailed work.

[23] B. Hofmeister, *Die Transhumance in den westlichen Vereinigten Staaten von Amerika,*" Ph.D. dissertation, Freien Univ. Berlin, 1958; C. L. White, "Transhumance in the Sheep Industry of the Salt Lake Region," *Economic Geography,* Vol. 2 (1926), 414–25; P. Scott, "Transhumance in Tasmania," *New Zealand Geographer,* Vol. 10 (1955), 153–72; P. Deffontaines, "Transhumance et mouvements de bétail en Amérique latine," *Cahiers d'outre mer,* Vol. 18 (1966), 258–94, 321–41.

economy; an advanced agricultural technology; and most importantly, a rural population that was strongly rationalistic and opportunistic in its attitude toward the land.

Yet Baker was not averse to qualifying his thesis of a finer adjustment of land use to physical differences. He noted that the influence of climate is more commonly exerted indirectly than directly. Thus, "the crop or farm enterprise most limited by climate, or otherwise, will, ordinarily, have the first choice of land, and will tend to push the less profitable crops beyond the periphery of its climatic limit, or at least keep them off the land necessary to produce a sufficient quantity to meet the effective demand." Nor did Baker foreclose the possibility that the trend toward closer adjustment to environmental variations might eventually be reversed if transportation became more expensive. In this case, he foresaw geographic differentiation in agricultural production persisting, but on a basis "more local than national or international in character—more like the agriculture of China."[24]

Baker's principal observation has since been corroborated in a wide variety of contexts. One of the most impressive of these subsequent illustrations is Prunty's work on the effects that the retreat of cotton to the best lands has had on the boundaries of the Cotton Belt. This contraction has proceeded so far, he reveals, that the Belt is now only a series of cotton regions whose over-all extent only dimly recalls the old east-west belt outline and whose total area comprises a definite minority of farmland in the South. At the same time, intensification of production on the better soils has not only increased total American cotton production, but has also fostered a shift in the median center of production precisely the opposite of the historic westward movement of cotton farming: eastward and across the Mississippi River.[25]

Yield intensification, areal specialization, and cropland contraction also have a growing part in the progressive blurring of distinctions between rural and urban populations. In his study of agricultural census data for the Northeastern United States, Hart noted the large areas that have been abandoned agriculturally but are becoming more important as residential and recreational areas for city people. At the same time, farming in the remaining croplands around the coastal urban masses is becoming much more specialized and much more productive per acre.[26] We may also note that the remaining farm population, being closer to cities and using ever more modern techniques, is rapidly approaching its urban counterpart in both residential and occupational characteristics.

Another aspect of increasingly better adjustment of land use to

[24] O. E. Baker, "The Increasing Importance of the Physical Conditions in Determining the Utilization of Land for Agricultural and Forest Production in the United States," *Annals of the Association of American Geographers,* Vol. 11 (1929), 17–46.

[25] M. Prunty, Jr., "Recent Quantitative Changes in the Cotton Regions of the Southeastern States," *Economic Geography,* Vol. 27 (1951), 189–208.

[26] J. F. Hart, "The Three R's of Rural Northeastern United States," *Canadian Geographer,* Vol. 7 (1963), 13–22.

environmental variations can be seen in the ways farmers have coped with increasing urban encroachment on farmland. In rapidly urbanizing southern California, farmers are, wherever possible, relinquishing marginal land first. This, plus the greater use of multiple cropping, more fertilization, switching to higher-value crops, and other intensification measures, has enabled the region to increase the value of its production from 1939 to 1959 by six times, despite a decline in crop acreage.[27]

European geographers have tended to view the contraction of agricultural land as a cultural loss. This is particularly evident in their study of the steadily lowering upper limits of cropping and summer pasturing in mountain areas. Although part of this lowering is due to the increased use of better lands at lower altitudes and nearer the farmstead, it is also often the result of the abandonment of farms as farmers leave the land for more lucrative work in the lowland factories. Not illogically, the French geographer, Blache, has labeled the decreasing intensification of mountain farming in Europe the *anticipation américaine,* using the Allegheny highlands as a sign of the future.[28]

Land economists have stressed the closer adjustment of land use to the environment as a prime example of the growing importance of technological advances and operational costs in modern agriculture. Andreae has demonstrated that the substitution of machinery for animal power has made distance from the farmstead less of an obstacle to the use of better-quality outlying plots and, at the same time, has discouraged the farming of hilly areas where the heavier machinery is handicapped and therefore less able to compensate for its high amortization costs. The cost of transportation is still an important factor in the farmer's bookkeeping, however, and where distances become great enough it can be at least as formidable as those agents encouraging farming specialization in the most favorable areas. One can see this well in Biehl's findings, which show that, although farmers in the Northeastern United States have succumbed to the competition of milder regions in certain commodities, they have maintained the total production of their area by capitalizing on their greater proximity to the large markets.[29] Here, too, one may see the eventuality posed by Baker, that as transportation costs increase, the trend toward closer adjustment to environmental differences might eventually be reversed.

Cultivation and Climatic Change

A third approach to the role of man-environment relationships in agricultural geography reflects a still further shift in interest from the natural environment to the human side of agricultural geography. The

[27] H. F. Gregor, "Urbanization of Southern California Agriculture," *Tijdschrift voor Economische en Sociale Geografie,* Vol. 54 (1963), 273–78.

[28] J. Blache, *L'homme et la montagne* (Paris: Gallimard, 1933), p. 162.

[29] B. Andreae, *Betriebsformen in der Landwirtschaft* (Stuttgart: Ulmer, 1964), p. 355; Biehl, *op. cit.*

emphasis here is not on the ways in which "agricultural man" works with his environment, but on the ways he is changing it. Like the "working-relationship" view of man-environment relationships, this one cannot be chronologically separated from the other views, even though research in this area is far more active now than it has been before, having at least as wide a representation of scholars from the various specialties as any of the other approaches.

Certainly the most controversial question pursued by this group has been whether man, in his agricultural operations, has materially changed regional climates. Somewhat surprisingly, more students from the physical sciences, in particular foresters and pedologists, claim that such changes have indeed been made and are impressive; geographers, on the other hand, have been relatively less impressed. But the geographers are by no means unanimous: more in Europe than in North America accept the view that there have been significant human alterations of climate. Fels aptly summarizes the affirmative view when he asserts the following effects on climate of a forest cover and thus, by implication, the reverse effects one might expect from its removal: cooler summers; smaller temperature variations; fewer frosts and of shorter duration; greater rainfall; higher humidity. The same effects, Fels maintains, can be expected where irrigation is introduced; the draining of lands, in contrast, would produce reverse effects, particularly on frost.[30]

Fels's résumé rests on a wealth of work going back well into the nineteenth century. Among Americans, the first pioneering study of man as the transformer of his landscape was by Marsh, originally published in 1864. Marsh touched upon a wide range of human modifications and their effects, of which climatic consequences were but one group. Although some of his statements have to be modified in the light of subsequent information, his view is upheld by a great number of modern scientists. Fels claims that the climate of ancient Germany was "markedly and measurably" different from today's and that this change can be attributed largely to extensive deforestation. Albrecht states that accelerated drainage through the deforestation and continued cultivation of the Eastern United States has drastically reduced the soil moisture supply, thereby also reducing the amount of water that is evaporated and thus lowering summer temperatures. Droughts are consequently becoming more severe and extensive. Curtis believes that the most convincing arguments for the influence on climate of the removal of vegetation are those connected with the energy balance as it is influenced by the albedo, or reflecting power of the earth's surface. Where grasslands on the northern Great Plains are plowed, Curtis has found that the dark prairie soils cut down reflection tremendously and accordingly absorb more solar energy—enough, he believes, to modify the temperature significantly. Alisov, Dresdov, and Rubenstein are typical of the Soviet scientists, who firmly

[30] E. Fels, *Der wirtschaftende Mensch als Gestalter der Erde* (Stuttgart: Franckh, 1954), pp. 124 f.

believe that measures such as planting forests and draining land will radically change regional climates.[31]

Opposed to most such views is Thornthwaite, who asserts that climate, by its own constant and often wide fluctuations, seriously limits any attempt to discover the effects of human influences upon it. Moreover, he attacks the thesis of those who claim that deforestation leads to less rainfall and that afforestation leads to the reverse. Although he admits that "a great mass of experimental evidence" demonstrates that various types of vegetation and land use do contribute varying amounts of water to the atmosphere, he also declares that there is no similar demonstration that the moisture added to the atmosphere in this way is reprecipitated later in the same area. Finally, he warns that observations made with more refined instruments and techniques of observation, which would better identify any climatic modifications, are still greatly lacking.[32]

As in all arguments, however, the differences between the two sets of opinions are not absolute. Many, in both groups, readily admit that atmospheric modifications have been made by man in his alteration of the vegetative cover; the debate revolves more around how much of the atmosphere is affected. Proponents of climatic change frequently use the word "local" and their opponents usually make their objections in the context of "micro-climates." Then, too, there has been a tendency to confuse the amount of area affected by the changes in land use with the amount of climatic change. In reality, only a few degrees change in temperature, for instance, and that for only a few hours, can result even from noteworthy expansions of cropland. This has been the observation of several geographers who have worked in Canada, all of whom have noted how the frequency of frost has been reduced by sizable clearing of land; yet none has claimed that this amelioration of temperature constituted a major climatic change.[33] Many others, to be sure, have so claimed.

A qualified agreement also exists about the extent of human modification of the climate by irrigation. This is the only method that Thornthwaite admits is capable of seriously changing a regional climate. But the changes he has in mind are those resulting from the effects of a changed soil moisture content on plant growth, whereas the other side

[31] G. P. Marsh, *Man and Nature: Or, Physical Geography as Modified by Human Action,* ed. D. Lowenthal (Cambridge, Mass.: Belknap Press of Harvard Univ., 1965).

Fels, *op. cit.,* p. 123; W. A. Albrecht, "Physical, Chemical, and Biochemical Changes in the Soil Community," in *Man's Role in Changing the Face of the Earth,* ed. W. L. Thomas, Jr. (Chicago: Univ. of Chicago, 1956), pp. 654–57; J. T. Curtis, "The Modification of Mid-latitude Grasslands and Forests by Man," *ibid.,* p. 733; B. P. Alisov, O. A. Dresdov, and E. S. Rubenstein, *Kurs Klimatologii* (Leningrad: Publishing Institute of Hydrometry, 1952).

[32] C. W. Thornthwaite, "Modification of Rural Microclimates," in *Man's Role in Changing the Face of the Earth,* p. 570.

[33] E. D. Albright, "Crop Growth in High Latitudes," *Geographical Review,* Vol. 23 (1933), 616; E. Ehlers, *Das nördliche Peace River Country: Alberta, Kanada* (Tübingen: Univ. Tübingen, 1965), p. 161; Deffontaines, *L'homme et l'hiver,* pp. 87 f.

thinks of these changes as being the direct effect of wetter surfaces, primarily reservoirs, on the atmosphere. Thornthwaite admits that such changes may be important to farmers located immediate to these water bodies, but Fels and his allies feel that the influence of the water is more extensive, particularly because of the many reservoirs and their scattered distribution. A review of the conflicting claims again shows that many of the arguments seem to assume that large areal influence is synonymous with major atmospheric influence.

There is complete and enthusiastic agreement, however, about the need for more and detailed devices and techniques of observation. Thornthwaite quotes Marsh extensively as cautioning against making too-broad conclusions on things climatic before adequate meteorological observations are made. Since that time, instruments and theory have improved to where it is now becoming possible to record even small changes occasioned by the planting of various crops. Not that this will automatically solve the generations-long debate. The high costs of instrumentation are still a barrier to adequate observation, and not all the methods of measurement are easy to understand or to use. The ultimate problem, however, may well be agreement on the statistical threshold at which a significant climatic change has been brought about.

Agricultural Landforms

Geographers are far more in agreement about the effects of man and his agricultural efforts on the configuration of the land. Although these evidences are only on the order of minor relief, they are definitely visible and in many areas quite extensive. The controversy is confined to the manner of human origin, although it has certainly been no less lively than that accompanying the topic of man-induced climatic change.

Ridged fields and terraces are the most prominent agricultural landforms, and each group has a variety of subforms. Ridged fields have a long ancestry, predating the plow. Evans believes this to be the case for those found in Ireland, Scotland, Norway, and Switzerland; Parsons, Bowen, Wilken, Eidt, and others reach the same conclusions about the ridged areas they have studied in Latin America. The majority of these fields are now abandoned, with many of the remaining areas still being worked with simple instruments. Drainage seems to have been a common motive for the construction of these ancient ridged fields, but not all investigators agree that it has always been the primary motivation. Evans believes that those in Europe were due even more to a wish to conserve fertility, the common practice being to pile the earth dug from the ditches onto the undisturbed sod of the intervening ridges, thereby enriching the organic material and preventing the plant nutrients from being washed down. Irrigation has been suggested as an additional, though secondary, reason for ridging fields in certain parts of Latin America. The suggestion seems logical in view of the many irrigated areas today that exhibit a ridged terrain, although these ridges are much more uniform and considerably narrower.[34]

Another type of ridged field, but in this case one that was originally plowed, is the distinctive "ridge and furrow" terrain found scattered through a band stretching from Germany to England. These ridges, or "arched fields" (Ger. *Wölbäcker*), were made as the farmer plowed the furrows toward the middle of his narrow strip of field, a common field form in these regions. Where plowing commenced in the center of the field and all subsequent turns were made within, rather than out of, the field, the ridge also developed an arched outline in longitudinal profile and an "S" shape in plan. Drainage has long been given as the main reason for ridge and furrow, but dissenters are not lacking. The fact that ridges and furrows often occur on lands not offering any drainage problem has convinced Jäger that their purpose can be discovered only by an investigation of the soil conditions of the particular area. Thus, where the terrain is flat and the soil light and dry, he would explain the ridges and furrows as a response to the need to conserve water. Ridges and furrows that trend without regard to slope he would attribute to the need for bringing less weathered soil to the surface of the fields.[35]

Kittler completely rejects these explanations and maintains that the basic reason for ridge and furrow terrain has been the small size and narrowness of the fields and the consequent desire of the farmer to prevent a loss of his land to his neighbors through erosion and cultivation. Plowing the furrows toward the middle of the field has seemed to be the best way to do it. Kittler also suggests that the close confines of the land inheritance system explain why this type of plowing (Eng. "gatherings"; Ger. *Zussamenackern;* Fr. *labour en adossant*) still continues on many small French and German farms despite the availability of the two-way or reversible plow, which makes such a procedure unnecessary.[36]

Kittler's argument would thus also seem a further complication to those investigators who see ridges and furrows as a possible evidence of past field patterns. If, as Kittler says, many of the ridges are still being formed, then their value as fossils is diminished considerably. On the other hand, if the ridges under study are at variance with present land use patterns, then his argument would favor those who claim that the pattern is a relic of medieval fields over those who feel that it is an indication of past drainage efforts, not necessarily indicative of any particular period. This last debate, also an old one, has been most recently

[34] E. E. Evans, "The Ecology of Peasant Life in Western Europe," in *Man's Role in Changing the Face of the Earth,* p. 233; J. Parsons and W. A. Bowen, "Ancient Ridged Fields of the San Jorge River Floodplain, Colombia," *Geographical Review,* Vol. 56 (1966), 317–43; G. C. Wilken, "Drained-Field Agriculture: An Intensive Farming System in Tlaxcola, Mexico," *ibid.,* Vol. 59 (1969), 215–41; R. C. Eidt, "Aboriginal Chibcha Settlement in Colombia," *Annals of the Association of American Geographers,* Vol. 49 (1959), 374–91.

[35] H. Jäger, "Die englische Kontroverse im Blickfeld der deutschen Siedlungsforschung," *Zeitschrift für Agrargeschichte und Agrarsoziologie,* Vol. 1 (1953), p. 21.

[36] G. A. Kittler, "Das Problem der Hochäcker," *ibid.,* Vol. 11 (1963), 157.

raised by Beresford and Kerridge in England, and its outcome is still unsettled.[37]

Although the ridges and furrows of Northwestern Europe seem to be peculiar to that region for historical and social reasons, one cannot be completely sure that similar configurations do not exist elsewhere. In North America especially, where fields have always been relatively large and farmers more able to purchase the latest equipment, including plows, from an ever-expanding store of farm machinery, the possibilities of the development of this kind of relief seem few. Rather, we might expect to see what Juillard has observed of the cultivation practices of large farmers in lower Alsace: they use the bigger dimensions of their fields to plow through both their length and width (*labour en plat*); they also extensively employ the reversible plow, which allows them to throw the furrow in one direction over the entire field. Surfaces are thereby flattened instead of ridged.[38] But this situation is not typical of the majority of the large Western fields. Crisscross plowing is practiced much less than commonly believed. More importantly, the fixed plow is still much more widely used than the reversible plow; thus, the larger the field, the more important it is to plow it lengthwise into strips (lands) in order to reduce the time that would otherwise be wasted by moving the plow from one end of the field to the other. The result is a series of alternating swells and swales, the higher lands being built by plowing the furrows toward the inside and the lower lands being excavated by plowing the furrows toward the outside (Eng. "casting"; Ger. *Auseinanderackern;* Fr. *labour en refendant*).

Presumably the pattern is reversed the succeeding year to keep the surface from getting any rougher, the furrows on the higher land (called the "back furrow" in America and the "ridge" in England) now being turned toward the outside, and the furrows on the lower land (the "dead furrow" in America and the "finish" in England) toward the inside. Yet although this reversal is strongly recommended by farming experts, no extensive studies of its extent have apparently yet been made. If one can judge from Kittler's observations of certain German farmers who were using modern one-way plows and presumably had comfortable fields, the chances of successful and consistent accomplishment of the procedure are poor. Kittler found that the farmers were unable to compute the exact plow depth needed to lower the surface to the level from which they had raised it the previous year; another complication—and one normal to farming operations—was the intervening additional plowings required for soil conditioning and crop care.[39] Only when many more studies of this kind are made will it be possible to determine whether farmers are actually developing a ridge and furrow terrain or simply maintaining a related pattern of shifting undulations.

[37] M. W. Beresford, "Ridge and Furrow and the Open Fields," *Economic History Review*, Vol. 1 (1948), 34–45; E. Kerridge, "Ridge and Furrow and Agrarian History," *ibid.*, Vol. 4 (1951), 14–36.

[38] A. G. Haudricourt and M. J. B. Delamarre, *L'homme et la charrue* (Paris: Gallimard, 1955), p. 433.

[39] Kittler, *op. cit.*, p. 151.

Still other ridge forms have been noted, but their origins have aroused little controversy. Perhaps this is one of the reasons why writings on these forms are meager, or at least do not often appear in those scholarly periodicals that have an international circulation. Derruau describes a type of ridge and furrow terrain developed by French and Belgian farmers for better drainage which, however, is a hindrance to farm machinery and thus is dying out. Similar drainage measures have been observed in the fields of lowland northwestern Germany, northern Italy, French Canada, and Mexico. Ridges also can develop along the smaller ends of strip fields where plowing on both sides is in the same direction. Material accumulates here when it is thrown to the outside of the field as the farmer turns the plow. In northwestern Europe, these ridges (Eng. "balks"; Ger. *Ackerberge;* Fr. *crêtes de labours*) are so prominent and old that they have attracted geographers such as Hartke with their value as indices of past field patterns. Surprisingly few researchers, however, have devoted their energies to analyzing the intricate and much more widespread pattern of ridges constructed around irrigated fields. Less directly related to the agricultural process, as Schaefer has shown, are those road or path embankments that may be reinforced in part by adjacent accumulations from the turning of the plow but owe most of their height to the progressive lowering of the adjoining fields and their more erodable surfaces.[40]

Terraces constitute the second of the two major groups of relief forms resulting from agriculture. Their striking appearance and widespread distribution, especially in China, the Middle East, the Mediterranean Basin, and the Andes, have evoked a wealth of theories about their origin, none of which has escaped criticism. The most obvious explanation, that terraces are primarily an agricultural response to rough lands, has been considerably weakened by the reports of Spencer and Hale, who noted that the distribution of terracing does not agree with the extent of agriculture in rough areas. This leaves us with the question of why some rough areas have not been terraced while others have. Physical conditions, other than the degree of slope, certainly have played their part. Despois gives examples of several places in the Mediterranean hill-and-mountain lands where argillaceous slopes, without resistance to erosion, have proved impossible to terrace.[41] The contrast in insolation from slope to slope, so well exemplified in the terraces of south-facing slopes in the Rhineland, is still another of numerous illustrations.

Yet there are also many other regions that do not have critical physical variations but still exhibit a varied terrace pattern. The question

[40] M. Derruau, *Précis de géographie humaine* (Paris: Armand Colin, 1961), pp. 161, 193; W. Hartke, "Über die 'Ackerberge' und ihre Bedeutung als Index für das Alter agrarlandschaftlicher Grenzen," *Zeitschrift für Agrargeschichte und Agrarsoziologie,* Vol. 2 (1954), 173–77; I. Schaefer, "Zur Terminologie der Kleinformen unseres Ackerlandes," *Petermanns geographische Mitteilungen,* Vol. 101 (1959), 199.

[41] J. E. Spencer and G. A. Hale, "The Origin, Nature and Distribution of Agricultural Terracing," *Pacific Viewpoint,* Vol. 2 (1961), 2; J. Despois, "Pour une étude de la culture en terrasses dans les pays méditerranéens," *Géographie et histoire agraires,* Annales de l'est, Mém. 21 (Nancy: Faculté des Lettres et des Sciences humaines, L'univ. de Nancy, 1959), p. 107.

has not lacked possible answers. Could the difference be due to variations in farm size, the small farmer having to farm more intensively than the large? Could it be due to differences in landholding systems, the tenant being less concerned with intensification than the owner-operator? Is it the result of differences in modes of transportation—that is, people accustomed to carrying things by themselves might adjust to terrace levels more easily than those accustomed to using beasts of burden? Perhaps, too, those accustomed to using teams of animals to till the land would be discouraged from building terraces because most of them are not wide enough to allow the turning of a team with a plow, while those depending on hand-cultivation tools such as the hoe or spade could cultivate terraces of any width. Or could terraces be directly related to the particular kinds of crops emphasized in the farming system, one system emphasizing crops for which terracing is particularly favorable? Is it primarily the differences in seasonal labor demands that promote areal differences in terracing, a long layoff encouraging terracing as a way of spreading out the work? Unfortunately none of these suggestions has been seriously investigated, even by the European geographers in their voluminous studies of Mediterranean agriculture. Terraces have been treated only incidentally in most geographic works.

The attention to changes in the world distribution pattern of terraces has been somewhat greater. The suggestions already given for present pattern variations that cannot be explained primarily by physical differences show that geographers are now more ready than they were to ascribe the current decline of terracing to cultural rather than physical factors such as soil erosion. The previous, and far longer, period of expansion of terraces over the earth, geographers have tended to leave to the cultural historians and anthropologists. Yet it has remained for geographers to pose two sharply conflicting theories about the manner of this expansion. Spencer and Hale believe that agricultural terracing originated in the Middle East and from there spread, with accompanying modifications, westward into the Mediterranean Basin and Northwestern Europe, southward into East Africa, and eastward into Southeast Asia and the islands of the Western Pacific. A secondary center is also assumed to have developed in Indochina, its influence diffusing northward into China, Japan, and Korea; westward to India, Ceylon, and across the Indian Ocean to Madagascar; southward to Indonesia; and eastward to the Philippines (Fig. 1).[42]

The diffusion of terracing as a cultural innovation is in effect denied by Kittler, however, when he asserts that terracing can occur wherever man cultivates slopes and thereby accelerates the ever-present tendency toward soil creep. The loosened materials gradually slide down the slope and collect against the unplowed edge of the field. Retaining walls he would explain as an attempt by the farmer to prevent the slumping of the slope from getting any worse. Kittler also feels that this explanation of terraces as an accidental byproduct of cultivation solves another prob-

[42] Spencer and Hale, *op. cit.*, pp. 30–38.

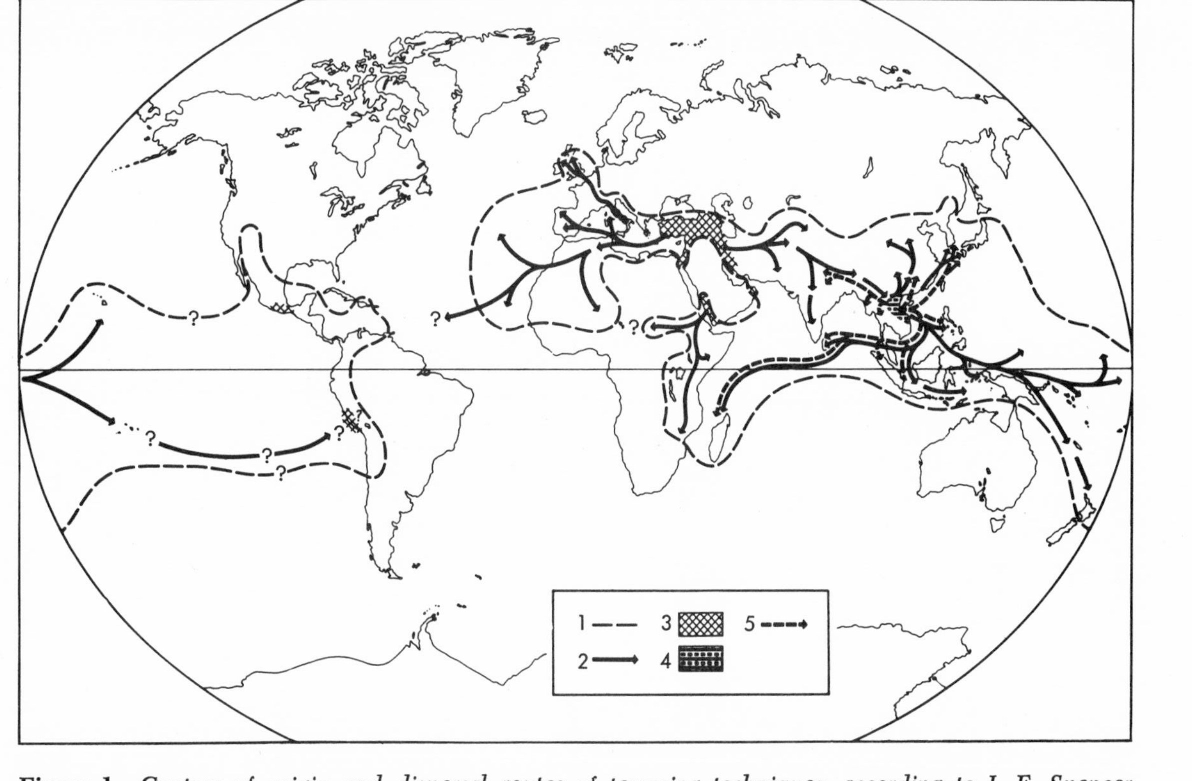

Figure 1. *Centers of origin and dispersal routes of terracing techniques, according to J. E. Spencer and G. A. Hale.* (1) *Outer limit of agricultural terracing;* (2) *Dispersal of earliest terracing technologies;* (3) *Region of origin of earliest terracing technologies;* (4) *Region of origin of wet field terracing;* (5) *Dispersal of wet field terracing. [By permission of the authors and* Pacific Viewpoint, *2 (1961), 33.]*

lem, the striking lack of written accounts of early terracing, only recently reemphasized by the difficulties encountered by Wheatley in his search of ancient Chinese literature on Indochinese farming for just such evidence.[43] Still a minority opinon, Kittler's is another step in a trend begun by several German geographers in the 1950's to assign more credit to physical processes than had previously been given in the explanation of the origin of terraces.

Whether terraces are a deliberate production by man or merely his unintentional acceleration of a geomorphic process is much more a point of argument among those who have studied agricultural terraces of a smaller order, particularly the strip lynchet (Ger. *Stufenrain;* Fr. *rideau*), which is widely scattered through Central and Western Europe.[44] The lower height and sod cover of its face, as well as its more inclined agricultural surface, have convinced many that the strip lynchet is the result of repeated plowings that have hastened the slumping of soil down slope (Fig. 2); others have maintained that the plows and plowing techniques being used would have made this slippage impossible but that the farmer deliberately encouraged it to form terraces and thereby prevent soil erosion.[45]

Like ridged fields, terraces also have a subgroup whose origin has involved little controversy. These also have less to do with the cultivation process, being constructed primarily for intercepting and retaining or diverting runoff at nonerosive velocities. The resultant profiles are thus exceedingly gentle, that of the first type (the level terrace) consisting of one broad ridge paralleled above and below by a similarly broad trough, and that of the second type (the graded-channel terrace) consisting of one ridge lying below a trough. Both types of terrace are widely distributed in the United States. Bennett believes that the graded-channel terrace may even have been developed and used most extensively in this country. The level terrace originated elsewhere, although its construction has been considerably improved by American effort. Bennett also partly echoes Kittler's theory when he claims that many "bench" terraces in the Southeastern United States today are former graded-channel terraces that have acquired a definite stepped appearance because of erosion encouraged by cultivation.[46]

Despite their extensiveness, ridged fields and terraces do not ex-

[43] G. A. Kittler, "Bodenfluss: eine von der Agrarmorphologie vernachlässigte Erscheinung," *Forschungen zur deutschen Landeskunde* (Bad Godesberg: Bundesanstalt für Landeskunde und Raumforschung), Vol. 143 (1963); P. Wheatley, "Agricultural Terracing," *Pacific Viewpoint,* Vol. 6 (1965), 123–44.

[44] G. Whittington, "The Distribution of Strip Lynchets," *Transactions, Institute of British Geographers,* Vol. 31 (1962), 128.

[45] Among those who believe the strip lynchet is the natural result of plowing are Kittler, *Bodenfluss;* and L. Aufrère, "Les rideaux: étude topographique," *Annales de Géographie,* Vol. 38 (1929), 529–60. Cf. M. P. Fénélon, "Les 'rideaux' de Picardie et de la Péninsule ibérique," *Bulletin de l'association de géographes français,* Nos. 255–56 (1956), 154; and G. Pfeifer, "The Quality of Peasant Living in Central Europe," in *Man's Role in Changing the Face of the Earth,* p. 251, who hold that it is a result of man's countermeasures against erosion.

[46] H. H. Bennett, *Soil Conservation* (New York: McGraw-Hill, 1939), pp. 443–76.

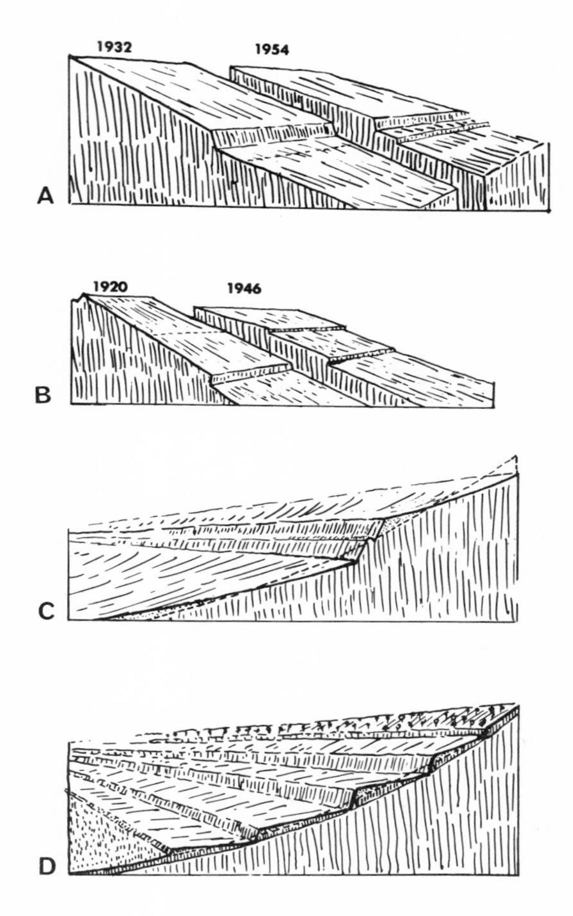

Figure 2. *Ways in which strip lynchets have developed through acceleration of soil creep, according to G. A. Kittler.* (A) *Plowing up to a fence located several feet down slope from a previously formed lynchet;* (B) *Cultivating along both sides of a garden path and accompanying retaining wall, constructed up slope from a previously formed lynchet;* (C) *Cultivating above and below a road;* (D) *Plowing above and below the grass borders of a field cultivated on the contour, the highest lynchet being only partially developed because of the greater resistance to soil creep provided by a tree crop. [By permission of the author and the* Institut für Landeskunde, Bad Godesberg. Forschungen zur deutschen Landeskunde, *143 (1963).]*

haust the types of relief of agricultural origin. Reclamation of former sea bottoms and coastal marshes have made available small, but highly productive, plains. Those along the North Sea coasts of the Netherlands and Germany have received the most attention. A less direct but no less impressive result of agricultural and other economic needs has been the rapid growth of canals and reservoirs. Noh has described the rich and intricate networks of irrigation canals in Japan, while Fels has devoted his attention to the nature and extent of the world's man-made lakes. Fels estimates that they now cover more than 193,000 square miles, a territory a fourth again as large as the State of California.[47]

[47] The classic study of reclamation in the North Sea area is P. Wagret, *Polderlands,* M. Sparks, tr. (London: Methuen, 1968). See also T. Noh, "On the Landscape of Irrigation Canals in Japan," *Comptes rendus du congrès international de géographie Amsterdam 1938,* Vol. 2, Sec. 5: *Paysage géographique* (Leiden: Brill, 1938), pp. 165–68; and E. Fels, "Die Stauanlagen und die Geographie," *Geographica Helvetica,* Vol. 20 (1965), 211; Fels, *Der wirtschaftende Mensch,* pp. 83–95.

The Economic Soil

The effects of agricultural man on the terrain are no more emphatic han the modifications he has made of soils. Traditionally, research has mphasized the damaging effects of man on soils—erosion and depletion—nd the consequent necessity for working in harmony with a more or ess inflexible nature. A more recent trend, however, is toward paying reater attention to the ways man has improved soils for his particular eeds, thus emphasizing man's independence of and superiority to a alleable environment. Pedologists, economists, and geographers have all ontributed to this view.

All these proponents have stressed the greatly increased producivity of soils resulting from human modification, but in a variety of ways. he most common approach, but less so that of geographers, has been he enumeration of the equivalents gained in cropland area by intensiication and the description of the contributory technological advances. Marbut produced such a survey of Western Europe as early as 1925, lthough he also believed that similar future increases in food there nd elsewhere would be unlikely because of the law of diminishing reurns and the smaller amount of good soil available for agricultural inensification. Harris made a similar study of the United States more reently, and his opinion of the potentialities of agricultural technology s much more optimistic.[48]

Another way of studying the effect of man on soil production, more opular among geographers, is to concentrate on specific soil regions. The study of the podzolics in the Northeastern United States has coninced Wolfanger that the striking contrast between their only moderate atural fertility and their premium productivity derives from their good tructure and their low content of plant food, both of which enable the oil type to be built up to meet the particular requirements of any numer of crops climatically tolerant of the region. Attitudes toward the proluctive capabilities of tropical soils are also beginning to change. Gourou as attacked the still commonly held view that volcanic soils are deerminants of high population density, by citing the high densities on the onvolcanic soils of the island of Madura, an accomplishment he atributes to artificial terracing, rotation, manuring, and other intensive gricultural practices.[49]

Associated with this regional approach to contrasts between natural ertility and soil productivity are the more ambitious attempts to map roductivity patterns on a world scale. Thoman, in his map of "Agricul-

[48] C. F. Marbut, "The Rise, Decline, and Revival of Malthusianism in Relaion to Geography and Character of Soils," *Annals of the Association of American Geographers,* Vol. 15 (1925), 1–29; C. D. Harris, "Agricultural Production in the United States: The Past Fifty Years and the Next," *Geographical Review,* Vol. 47 1957), 175–93.

[49] L. A. Wolfanger, "Economic Geography of the Gray-Brownerths of the Eastern United States," *ibid.,* Vol. 21 (1931), 277; P. Gourou, "The Quality of Land Use of Tropical Cultivators," in *Man's Role in Changing the Face of the Earth,* . 343.

tural Productivity of the World's Soils," divides the land area into thre productivity categories: "high," "intermediate to low," and "very low. Only the Northeastern United States, Northwestern and Central Europe the black soil belt of the Soviet Union, Japan, and certain desert oase are classified as heavily producing regions. A similar map has been draw by the Woytinskys, although it is more generous in its delimitation of th "high" regions. These areas include, besides those shown on Thoman' map, the majority of lowland Southeast Asia, the southern tips of Aus tralia and the enclaves of the humid Pampa and Rio de Janeiro–Sã Paulo areas in South America. In somewhat the same vein, Gregor ha classified each of the principal world soil types as either a "better eco nomic soil" or "poorer economic soil," using the criteria of both presen and potential productivity. Prairie, podzolic, and alluvial soils compris the most favored regions. Any future attempts at this last type of classi fication promise to be much easier, now that the U.S. Soil Conservatio Service has developed a new soil classification system, called "Soi Classification, Seventh Approximation." Now soils will be categorize according to properties that are the result of man's activities instead o on the basis of virgin-soil properties.[50]

[50] R. S. Thoman, *The Geography of Economic Activity* (New York: McGraw Hill, 1962), p. 118; W. S. and E. S. Woytinsky, *World Population and Production Trends and Outlook* (New York: Twentieth Century Fund, 1953), p. 319; H. F Gregor, *Environment and Economic Life* (Princeton, N.J.: Van Nostrand, 1963) pp. 208–10; C. E. Kellogg, "Why a New System of Soil Classification," *Soil Science* Vol. 96 (1963), 1–5; R. W. Peplies, "The New Soil Classification System and Geo graphic Methodology," *Southeastern Geographer,* Vol. 3 (1963), 1–8.

CHAPTER FOUR

Spatial organization

The Thünen Theory in the Landscape

The rising interest in the role of man as creator and changer of the gricultural landscape is also reflected in the greater attention being iven to the importance of human motivations to this transformation. conomic, cultural, and political influences are of course inseparable in heir operations, but this has not deterred geographers and other speialists from attempting to isolate these influences and to weigh their elative importance.

The research into economic motives has been unique in that so nuch of the work revolves around the ideas of one man, Johann Heinrich on Thünen, who lived in northern Germany from 1783 to 1850 and was oth an agricultural economist and a practicing farmer. Thünen was conerned with the kind of distributional patterns of farming systems that vould develop around an urban market if one isolated the variable of ransportation cost and held all other factors constant. He therefore asumed: (1) a uniform physical environment; (2) a completely commerial economy in which the farmer was both desirous and capable of naximizing his profit; (3) only one means of land transportation and hat with costs directly proportional to distance; and (4) an area in which oth market and hinterland were solely dependent on each other for heir existence (the "isolated state," as Thünen expressed it). Thünen hen proceeded to demonstrate that farmers would farm less intensively vith increasing distance from the market for, as their transportation costs ncreased, the operators would have to compensate by reducing their nvestment. The resulting land use pattern, therefore, would become one f concentric zones or belts surrounding the market, the outer boundary f each zone being determined by the distance at which farmers prac-

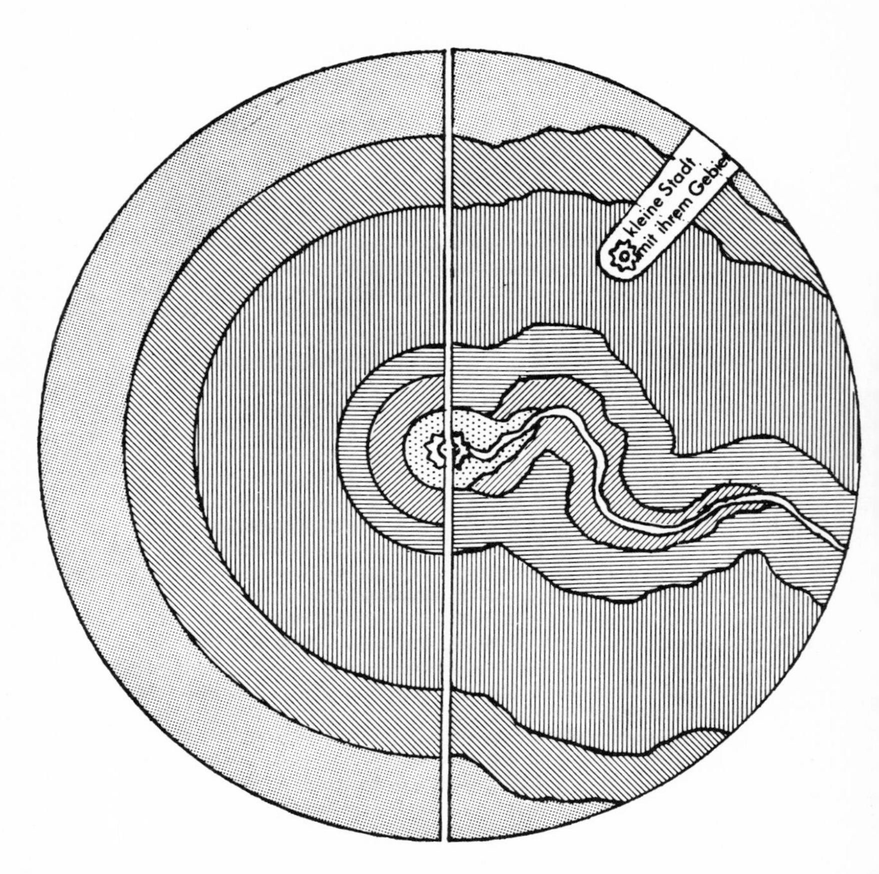

Figure 3. *The Thünen zones and their modification by a river (as they appeared in Thünen's work of 1826). Beginning with the dotted pattern at the center and proceeding outward:* (1) *Market gardening;* (2) *Forestry;* (3) *Intensive crop farming (grazing and root crops);* (4) *Crop farming (with fallow and pasture);* (5) *Three-field rotation;* (6) *Grazing. At the lower right appears "a small city with its hinterland." [By permission of H. B. Johnson; reproduced from the* Annals of the Association of American Geographers, *52 (1962), 216.]*

ticing a particular farming system could no longer make a profit because of increased transportation costs and thus would have to switch to a more extensive system (Fig. 3). The work of Thünen thus represents the first detailed exposition of regularity in the spatial arrangement of agricultural types, the relative superiority of one or the other farming system, and the ability of agriculture in similar physical environments to develop completely different forms.

Despite these contributions, Thünen's theory received little attention for many years after its first publication, in 1826. Not until around the turn of the century did agricultural economists, such as Engelbrecht and Krzymowski, earnestly begin to evaluate the work and to make it an integral part of their field. The year 1925 seems to mark the first serious consideration of the theory by German geographers, such as Schmidt and Sapper.[1]

Among English-speaking researchers the situation was no better, if not worse. One of the first treatments of the theory by agricultural economics was Ely and Wehrwein's, in 1940, although the English translations of earlier studies of the theory by Brinkmann and Krzymowski have been much more influential. Jonasson's 1925 article on European agriculture appears to be the first full-scale presentation of Thünen's theory in an English-language geographical journal, but the most detailed descriptions by English-speaking geographers—Grotewold, Chisholm, Johnson, and Hall (and his translator, Wartenberg)—did not begin to appear for another thirty years. That Hall and Wartenberg's work is the first English translation of Thünen's "Isolated State" points up one of the reasons for its long neglect: German syntax is difficult, and Thünen's is involved even for German readers. Also, Thünen's work was done when few agricultural economists were as interested in the theoretical aspects of agriculture as they were in the immediate problems of the field, a situation that still exists to a significant degree among American rural economists.[2]

Evidences of the Thünen circular pattern in the landscape, however, have been reported by writers for over a century. One of the few who paid more than passing attention to Thünen's writings during the long period of their neglect, Roscher, affirmed the "statistical reality" of the belts around London in 1845 and around Capetown in 1859. Similar descriptions are still being made. Using data about production and the area of crops in 1950, Stamer has claimed the existence of an irregular concentric zoning of land use around Hamburg, with allotments (*Schrebergärten*) nearest the city; commercial horticulture, fruit, osiers, and nurseries farther out; and field-crop and livestock farming still more distant. Buchanan and Hurwitz have distinguished a coarse four-zone gradation from the South African port city of Durban to the interior: horticulture to dairying to cereals to meat. A five-zone gradation was

[1] J. H. v. Thünen, *Der isolierte Staat in Beziehung auf Landwirtschaft und Nationalökonomie* (Hamburg: Perthes, 1826).

T. H. Engelbrecht, "Der Standort der Landwirtschaftszweige in Nord-Amerika," *Landwirtschaftliche Jahrbücher,* Vol. 12 (1883), 459–509; R. Krzymowski, "Bemerkungen zur Thünenschen Intensitätstheorie und ihrer Literature," *Fühlings Landwirtschaftliche Zeitung,* Vol. 50 (1901); P. H. Schmidt, *Wirtschaftsforschung und Geographie* (Jena: Fischer, 1925), pp. 68–70; K. Sapper, *Allgemeine Wirtschafts- und Verkehrsgeographie* (Berlin: Teubner, 1925), pp. 159 f.

[2] R. T. Ely and G. S. Wehrwein, *Land Economics* (New York: Macmillan, 1940); T. Brinkmann, *Theodor Brinkmann's Economics of the Farm Business* (Berkeley: Univ. of California, 1935); R. K. Krzymowski, "Graphical Presentation of Thuenen's Theory of Intensity," *Journal of Farm Economics,* Vol. 10 (1928), 461–82.

O. Jonasson, "Agricultural Regions of Europe," *Economic Geography,* Vol. 1 (1925), 277–314; A. Grotewold, "Von Thünen in Retrospect," *ibid.,* Vol. 35 (1959), 346–55; M. Chisholm, *Rural Settlement and Land Use* (London: Hutchinson, 1962); H. B. Johnson, "A Note on Thünen's Circles," *Annals of the Association of American Geographers,* Vol. 52 (1962), 213–20; P. Hall, ed., C. M. Wartenberg, tr., *Von Thünen's Isolated State* (Oxford: Pergamon, 1966).

observed by Kühn south of Buenos Aires, comprising horticulture at the city margin; rotation of wheat, corn, and flax and intensive livestock farming at moderate distances from the city; more extensive field-crop farming still farther out; and extensive sheep raising in Patagonia. In 1958, Andreae schematized a similar number of zones of decreasing farm intensity in a southwesterly direction from Chicago: (1) fruit, vegetables, fresh milk, "industrial finishing"; (2) cash grain crops (corn, soybeans, oats); (3) manufacturing milk, grain-hog farming; (4) grain-hog and grain-cattle farming; and (5) pasture-livestock farming.[3]

Investigators have also perceived resemblances to the Thünen pattern on a continental and even world scale. Jonasson presented a detailed zonation of Europe in 1925. He postulated five production belts, several of which were further subdivided into zones. Nearest the cities in the Northwest, horticulture dominated, with greenhouses and floriculture in zone 1 and truck products, fruits, potatoes, and tobacco in the outer zone 2. Next came intensive agriculture and intensive dairying, which included zone 3 (dairy products, beef cattle, mutton sheep, veal, forage crops, oats, flax for fiber) and zone 4 (general farming; grain, hay, livestock). Extensive agriculture was next, concentrating on bread cereals and flax for oil, followed by "extensive pasture" on which beef and range cattle, range horses, and range sheep were grazed. Finally, came the "outermost peripheral area," which was largely forest. The most ambitious presentation, however, was an earlier effort (1920) by Laur. He expanded the zonation beyond Europe and over the rest of the earth, arriving at seven major zones: residential, local, industrial, plantation, agrarian, pasture, and caravan. Each of these he then further divided into anywhere from three to seven sub-areas, or thirty-two farming systems in all.[4]

Much less thought has been devoted to application of Thünen's theory to the individual farm, although Thünen declared that the correlation between farming intensity and distance was just as valid for the smaller unit. To reduce the greater time and effort lost in moving back and forth between the farm buildings and the more distant fields, the farmer would have to use those areas less intensively. In 1936, Müller-Wille illustrated this adjustment by superimposing on a map of a German farm village and its surrounding lands a pattern of "isochrones," lines

[3] W. Roscher, "Das von Thünensche Gesetz," in *Archiv der politischen Oekonomie und Polizeiwissenschaft*, ed. Hermann *et al.*, Vol. 3: *Ideen zur Politik und Statistik der Ackerbausysteme* (Heidelberg: Winter, 1845), pp. 186–95; *idem, Nationalökonomik des Ackerbaus und der verwandten Urproductionen* (Stuttgart: Cotta, 1859), 187.

H. Stamer, "Die Standortsorientierung der Landwirtschaft um den Grossmarkt Hamburg," *Berichte über Landwirtschaft*, Vol. 33 (1955), 408–34; K. Buchanan and N. Hurwitz, "Land Use in Natal," *Economic Geography*, Vol. 27 (1951), 236; F. H. Kühn, *Das neue Argentinien*, cited by E. Obst, *Allgemeine Wirtschafts- und Verkehrsgeographie* (Berlin: de Gruyter, 1959), p. 234; *B. Andreae, Betriebsvereinfachung in der Landwirtschaft*, Berichte über Landwirtschaft. Neue Folge. 169. Sonderheft (Hamburg: Parey, 1958), p. 23.

[4] Jonasson. *op. cit;* E. Laur, *Einführung in die Wirtschaftslehre des Landbaus* (Berlin: Parey, 1920).

howing zones of equal time required in carting manure to the fields.[5] Vhere the isochrones spread out more, there also extended the farming ystem; and, of the two principal farming systems practiced by the armers, the most extensive was also the farthest away from the settle- nent.[6] Since World War II, as Europeans have become increasingly oncerned with the inefficiencies of the small farm and its fragmented ields, numerous studies of the relations between various economic in- lices and the distance from the farm settlement leave no doubt about the eed for consolidation of property, to eliminate much wasteful travel nd thus encourage more intensive over-all agriculture in the region. Evidences of zones of intensity around farms are also being uncovered n non-European areas, such as Central Africa and India.[7]

Thünen also proposed sectorial fields as the most rational means of educing the time spent by the farmer in traveling around the farm. In his way easy access to all fields, i.e., all types of land use on the farm, vould be guaranteed. Travel to the outer parts of the sectors also would no longer be wasteful, for the farmer could carry on his operations during he entire trip, as in plowing. For farmers in group settlements, another dvantage would be the possibility of combining the social amenities of he village with closeness to their fields. Though sectorial fields are not common, they are used in widely varying farming systems and environ- nents. Best known are the "radial forest farms" (Ger. *Radialwaldhufen*), ormed by certain groups of North European peasants as they cleared he land beyond their villages. These wheel-like openings mark an outer edge of farm settlement established during the Middle Ages.[8] One of the best examples of radial farm patterns of modern origin is the cooperative arm of Israel. A radial corral arrangement, or "pie" farmstead, is used by some dairy farmers in southern California to reduce the time and ravel needed for moving cows between the barn and corrals.[9] Tribal agriculturists in northern Ghana, in contrast, maintain an apparently well- established tradition by dividing their farms radially to ensure an equal sharing by the inheritors of the better-manured land near the house.[10]

Modifications and Contradictions

This impressive list of affirmations of Thünen's theory is more than matched by accounts of deviations from the theoretical pattern. Nor is this surprising, since Thünen never assumed that his "isolated state" was anything more than hypothetical or that it was designed to illustrate the locational significance of anything but transportation costs. Further-

5 Cited by E. Otremba, *Allgemeine Agrar- und Industriegeographie,* 2d ed. (Stuttgart: Franckh, 1960), p. 130.

6 Chisholm, *op. cit.,* pp. 50–73.

7 E.g., J. M. Hunter, "The Social Roots of Dispersed Settlement in Northern Ghana," *Annals of the Association of American Geographers,* Vol. 57 (1967), 341 f.

8 G. Schwarz, *Allgemeine Siedlungsgeographie* (Berlin: de Gruyter, 1961), p. 167.

9 H. F. Gregor, "Industrialized Drylot Dairying: An Overview," *Economic Geography,* Vol. 39 (1963), 308 f.

10 Hunter, *op. cit.,* pp. 346–48.

more, Thünen believed that his theory could serve as a tool for isolating the effects of other variables on an "ideal" pattern of land use based on transportation costs. In this objective he succeeded admirably, although the growing number of studies still show no universal agreement over just how strongly reality has altered the circular zonation or whether modifications do, in fact, always indicate a weakening of the influence of distance.

Several forms of real-life modifications of Thünen's zones have been advanced by researchers: centrifugal and centripetal movements; coalescence; distortion and fragmentation; and most recently, reversal of intensity gradation. The outward expansion of the zones as transportation improves and the city expands was described by Sax in 1878, in a study of the effects of railroads; since then, the same phenomenon has continued to be observed from various angles. Some investigators have described the outward movement of the over-all pattern of land use around a particular city, whereas others have concentrated more on just one part of the pattern, such as the shifts in dairy production.[11]

Another modification of the zones, coalescence, was first seriously treated by Engelbrecht in 1884, when he observed how the outer and more extensive zone of livestock raising was checked in its peripheral movement and was gradually being overtaken by more intensive crop zones. This expansion of the more intensive belts at the expense of the less intensive, reinforced by the same process around other markets, has been seized upon by some as evidence of an eventual disappearance of the spatial zonation of agricultural land use based upon market distance. Others, however, have used this same evidence as support for their position that the coalescence of individual zonal patterns merely leads to a broader pattern, of worldwide scope—the "isolated world state" whose center of intensity is Northwestern Europe. Engelbrecht, again, was one of the pioneers in this argument, and Obst is a recent advocate.[12]

Even this more general view of Thünen's theory has not escaped criticism, since the more extensive zones can move inward at the same time that the more intensive zones expand outward. Birch ascribes this outer retraction to the independent effect of technically induced increases in productivity in the intensive zones, and cites the United States and the United Kingdom as prime examples. A more fundamental objection to the concept of an isolated world state, however, has been advanced by Otremba. He notes that even if major disruptive effects of noneconomic character, such as environmental differences and variations in stage of cultural development, are ignored, it is still impossible to arrive at a

[11] E. Sax, *Die Verkehrsmittel in Volks- und Staatswirtschaft* (Vienna: Hölder, 1878). H. F. Gregor, in "Urbanization of Southern California Agriculture," *Tijdschrift voor Economische en Sociale Geografie*, Vol. 54 (1963), 273–78, has described the outward movement around Los Angeles, e.g., and L. Durand, Jr., has reported on the shifts in dairy production in "The Major Milksheds of the Northeastern Quarter of the United States," *Economic Geography*, Vol. 40 (1964), 9–33.

[12] T. H. Engelbrecht, "Die Transportkosten auf Eisenbahnen und deren Einflüsse auf die Landwirtschaft," *Fühlings Landwirtschaftliche Zeitung*, Vol. 13 (1884), 533 f.; Obst, *op. cit.*, pp. 232–35.

single-intensity pattern because the patterns of maximum application of land, labor, and capital are not the same.[13] The principal center of land use intensity, measured in yield, is in Northwestern Europe; the area of highest capital intensity is divided among the American Middle West and our irrigated Western oases; the greatest labor intensity is in one of three places, depending on its definition—in Southeast Asia, if manual labor; in Northwestern Europe, if a combination of manual labor and low labor costs; in the United States, if mechanized labor. Nor is there a single worldwide pattern of intensity based on net returns, for the same gains can be made by different combinations of production factors. Net gain is high in the United States because a high intensity of capital easily compensates for a low yield, but net gain is also great in Northwestern Europe because its high intensity of labor and low labor costs more than counteract the lower intensity of capital.

Another modification of Thünen's scheme, related to centrifugal movement of the zones, is distortion along transportation lines. Outward movement on transportation routes is naturally faster than in the spaces between routes, so that outward-extending prongs and sectors are characteristic. Irregular spacing and variations in the lengths of these extensions are also to be expected, in view of the large investment required and the dependence of the transport route on terrain, climate, distribution of traffic, and other variables. Nor can the lengths of these outward extensions be expected to be directly proportional to distance. Alexander has shown that freight rates in the United States increase at a decreasing rate (the "tapering principle"), and Dunn points out quite logically that the mode of transportation as well as distance is an important determinant of transport cost.[14]

Because these elongations are most evident along the very edge of the agricultural ecumene, special attention has been accorded locations where railroads have penetrated the virgin areas. Jefferson provided some of the first continental views of these penetrations in his maps of "railwebs," and Mackintosh added historical perspective with his maps of railroad expansion on the Canadian prairies. Closer views of these extensions of the ecumene have shown that a well-defined sequence of gradation in land use can often be recognized, changes occurring in both the types and the areal proportions of crops as distance from the railroad increases. More recent evidence of this same phenomenon is offered in studies of the northern Peace River area, by Lovering and Ehlers, and, with respect to roads, in a work on the vicinity of Addis Ababa by Horvath. On the other hand, variations in the physical environment can modify such zonations to the point where the agricultural extension is only a scattering of cropland along a route, as Darby has shown in his

[13] J. W. Birch, "Rural Land Use and Location Theory: A Review," *Economic Geography*, Vol. 39 (1963), 276; Otremba, *op. cit.*, pp. 128–30.

[14] J. W. Alexander, S. E. Brown, and R. E. Dahlberg, "Freight Rates: Selected Aspects of Uniform and Nodal Regions," *Economic Geography*, Vol. 34 (1958), 1–18; E. S. Dunn, Jr., *The Location of Agricultural Production* (Gainesville: Univ. of Florida, 1954), pp. 64 f.

correlation of soil differences and farm clusters in the "railway belt" of former Northern Rhodesia.[15]

In more hospitable environments, where farming is usually more varied and intensive, distortion of the Thünen pattern is likewise more advanced. Durand has devoted particular attention to these complexities in his studies of fluid milk production in the vicinities of cities in the Northeastern United States. Most of the milksheds are highly irregular, he has found, and even in those that do approach the theoretical circular form certain segments of the "ring" are more productive than others. A few milksheds are ovular. Many also contain outlying, noncontiguous producing areas. Much of this irregularity in shape can be traced to the different and changing strengths of the various markets. As the milkshed of one city expands at the expense of a tributary to a smaller center, the smaller milkshed similarly compensates by expanding into areas of less resistance, i.e., areas where milksheds are still smaller. The expansion of the larger milksheds is also often accompanied by the establishment of outliers, so that where the disparity in the competitive power of two milksheds is especially great, the market in the smaller milkshed may be surrounded by areas producing milk for the larger market. Competition with other types of farming is another important determinant of milkshed shape, as Durand has observed along the boundary between the Dairy Belt and Corn Belt just south of Chicago. Despite being closer to the Chicago market than farmers farther north, along the Wisconsin-Illinois border, and despite having equally favorable railroad facilities, farmers in the Corn Belt section have refused to shift to dairying because the cash-grain or corn-hog-beef cattle farming system is more remunerative and requires less, and less regular, work. Thus, the Chicago milkshed has had to expand farther north into Wisconsin to compensate for this barrier.[16]

Regional specialization has become an increasing impediment to the development of the Thünen pattern as transportation has grown and improved, lowering transportation costs and thereby reducing the economic advantage of proximity to markets relative to other locational forces—principally climate, soil, and terrain. This trend toward areal specialization based more on environment than market distance is reinforced by technological and marketing changes that stress economies of scale.

The already impressive extent of regional specialization in such countries as the United States and Canada, and its growing spread into

[15] M. Jefferson, "The Civilizing Rails," *Economic Geography,* Vol. 4 (1928), 217–31; W. A. Mackintosh, *Prairie Settlement: The Geographical Setting,* Vol. 1 of *Canadian Frontiers of Settlement,* ed. *idem* and W. L. G. Joerg (Toronto: Macmillan, 1934), pp. 48–51; J. H. Lovering, "Agricultural Land Use in the Fort Vermilion-La Crète Area of Alberta," *Geographical Bulletin,* Vol. 20 (1963), 39–57; E. Ehlers, *Das nördliche Peace River Country, Alberta, Kanada* (Tübingen: Univ. Tübingen, 1965), pp. 167–71; R. J. Horvath, "Von Thünen's Isolated State and the Area around Addis Ababa, Ethiopia," *Annals of the Association of American Geographers,* Vol. 59 (1969), 308–23; H. C. Darby, "Settlement in Northern Rhodesia," *Geographical Review,* Vol. 21 (1931), 559–73.

[16] Durand, *op cit.,* p. 29; *idem,* "Dairy Region of Southeastern Wisconsin and Northeastern Illinois," *Economic Geography,* Vol. 16 (1940), 423.

other areas, like Western Europe, make it more difficult than ever to determine just how important distance still is as a location factor. American geographers, beginning with the masterly treatment of the process of regional specialization by Baker in 1921, have seen little in their intensely developed agricultural landscape to overly impress them with Thünen's theory.[17] Only the market milk-cheese-butter zonation from east to west in the Dairy Belt has been acknowledged as an extensive example of the influence of distance, and this pattern too seems to be dissolving.[18] Since World War II, the westernmost part of the Dairy Belt, the one traditionally specializing in manufacturing milk because it was farthest from urban markets, has begun to shift toward fluid milk to satisfy the growing demands of cities all over the Eastern United States. Further complicating the boundary between fluid and manufacturing milk areas has been the rapid and widespread introduction of the multiple-purpose plant which is able to manufacture butter or cheese, or ship fluid milk, as the market demands.

Europeans, in contrast, have been somewhat more impressed. The German agricultural economist, Andreae, besides outlining a five-ring zonation of livestock systems in the American Corn Belt, also claims evidence for a similar zonation extending over the Eastern United States and Western Europe. Yet he also notes that this picture is blurred in Europe because of the "overlapping influences of the individual location factors" and admits that there is a reinforcement of economic zonation by a similar zonation of environmental conditions. The British geographer Chisholm tackles the much more difficult problem of ascertaining the role of distance in the distribution of each of twenty-nine major crops raised in England and Wales. By comparing the share that each county has of the national acreage of each crop with the share that the county has of the total acreage of all twenty-nine crops, Chisholm has determined the amount of deviation from the national pattern, or the localization, of each of the crops. Despite some patterns that could not be explained satisfactorily, Chisholm believes he has found enough significant correlations between localization and markets to "manifestly reveal the importance of distance and perishability." However, he owns that it is hard to forecast how long this situation will last.[19]

The difference between the general American and European view of the pertinency of Thünen's concentric zones also shows up in recent studies of the effects of national trade policies on world intensity patterns. The American geographer Grotewold and his student Sublett dispute the assumption that the high agricultural intensity in all countries of North-

[17] O. E. Baker, "The Increasing Importance of the Physical Conditions in Determining the Utilization of Land for Agricultural and Forest Production in the United States," *Annals of the Association of American Geographers*, Vol. 11 (1921), 17–46.

[18] L. Durand, Jr., "Recent Market Orientations of the American Dairy Region," *Economic Geography*, Vol. 23 (1947), 32–40; G. R. Lewthwaite, "Wisconsin Cheese and Farm Type: A Locational Hypothesis," *ibid.*, Vol. 40 (1964), 95–112.

[19] Andreae, *op. cit.*, p. 23; *idem, Betriebsformen in der Landwirtschaft* (Stuttgart: Ulmer, 1964), pp. 143, 145; Chisholm, *op. cit.*, pp. 92–97.

western Europe is primarily the result of proximity to market, pointing out that farming in West Germany is much more intensive than it is in the United Kingdom, and this despite broad similarities in size, physical character, the size of the population, incomes, and tastes. This difference in agricultural intensity they attribute to stricter regulations of agricultural imports by West Germany. Chisholm, contrarily, carries on in the tradition of Engelbrecht and Laur, professing to see a worldwide pattern of zonation in the origins of the horticultural and dairy products imported by the United Kingdom. He also observes that much of the imports come into the country at precisely the time of year when tariffs are adjusted to give maximum protection.[20]

The few studies made of the deviations from the Thünen pattern on the individual farm present much the same picture: a decline in the influence of transportation relative to other factors. From a pattern of intensity that was dictated primarily by the distance between field and farmstead has evolved another that reflects more the role of the physical environment. Tractors and other farm machinery now allow the most level land to be cultivated easily, no matter how far it lies from the farmstead; costs, in fact, dictate that only the best land be cultivated. As more hired labor is used and more emphasis is put on crops that do not require the intermediate role of the farm complex, this loosening of the influence of the farmstead's location on the surrounding land use pattern is accelerated. Schroeder illustrates this well in his comparison of dairy farms, which have their grazing and part of their alfalfa areas closest to the buildings, and crop farms which have more lax spatial arrangements.[21]

Improved transportation also appears to be the reason for the increasingly eccentric position of the farmstead among its holdings, in this case because of an even further tightening, not loosening, of transportation bonds. Smith has noted this trend in the United States where, as soon as a new highway is completed, farmers begin relocating their houses and barns in order to be on it. A further stimulus to orientation of settlement to the road has been the increase in farm size with the consolidation of smaller farms, noted by Kollmorgen and Jenks. One may even doubt whether most American farmsteads have ever been located very far from the highway in the first place. The postulate that isolated farmsteads locate at the central location within the farm or at the point which minimizes travel to perform all farming operations is not supported by at least the more detailed farmstead studies.[22]

[20] A. Grotewold and M. D. Sublett, "The Effect of Import Restrictions on Land Use: The United Kingdom Compared with West Germany," *Economic Geography,* Vol. 43 (1967), 64–70; Chisholm, *op. cit.,* pp. 97–105.

[21] K. Schroeder, *Agrarlandschaftsstudien in südlichsten Texas* (Frankfurt am Main: Kramer, 1962), p. 127.

[22] T. L. Smith, *The Sociology of Rural Life* (New York: Harper, 1953), p. 273; W. M. Kollmorgen and G. F. Jenks, "A Geographic Study of Population and Settlement Changes in Sherman County, Kansas," *Transactions, Kansas Academy of Sciences,* Vol. 54 (1951), 449–54. A recent detailed study of farmsteads is I. Burton, *Types of Agricultural Occupance of Flood Plains in the United States* (Chicago: Univ. of Chicago, Dept. of Geography, 1962), p. 26.

Transportation changes have obviously little affected the layout of farmstead buildings, although centrality of functions is now being increasingly emphasized in the time-and-motion studies of agricultural economists. Their recommendations, the growing obsolescence and disrepair of buildings, as well as the continuing urge to rationalize farm operations in order to fully realize the benefits of newer farm technologies, would seem to point to a more transportation-oriented farmstead plan than before. The radial arrangement of corrals around the milking parlor on some of the newer dairy farmsteads in California and Arizona is an example of this trend.[23]

Quite different has been the trend in relationships between transportation changes and the zonation of the intensity of land use near cities. Geographers and economists generally agree that the dense transportation nets of the heavily urbanized parts of Northwestern Europe and Eastern North America have now eliminated practically all traces of a Thünen pattern. Until recently, they also agreed that most of these remnants were to be found in a zone of intensive farming on the urban fringe; now, several North American geographers believe that even this innermost zone of intensity is disappearing and that farming there is actually becoming more extensive than that farther out. This inverted zonation they attribute to increased competition from more distant areas, but even more to the rising cost of land near the cities as a result of the real estate boom. Grotewold was the first to propose this zonation reversal, and Sinclair has offered a ring-type model (Fig. 4). The progression of the intensity of Sinclair's rings is directly proportional to the degree of urban influence in the forms of high urban taxes, restrictive zoning, and nuisances stemming from nearby urban areas. More recently, two English writers, Best and Gasson, have added their support to this theory in their study of a trend toward extensiveness along the southeastern margins of London. They believe that the shift is largely the result of: (1) increasing competition from areas more distant but with otherwise better production facilities; and (2) the loss of casual labor to city jobs by urban-fringe farmers.[24]

The view that the decreasing intensity of farming on the urban fringe is anything more than a temporary and probably accidental event contradicts a long record of opinion. As early as 1914, Brinkmann declared that the innermost zone of intensity could not disappear because of lowering transportation costs so long as the market continued to expand in population and purchasing power. In 1936, Pfeifer felt that he had uncovered enough evidence in a study of the predominantly export-oriented agriculture of central and northern California to prove that a city can still exert a localizing effect on the zonation of intensity, even

[23] Gregor, "Industrialized Drylot Dairying," pp. 308–10.

[24] Grotewold, *op. cit.*, pp. 346–55; R. Sinclair, "Von Thünen and Urban Sprawl," *Annals of the Association of American Geographers*, Vol. 57 (1967), 72–87; R. H. Best and R. M. Gasson, *The Changing Location of Intensive Crops*, and R. M. Gasson, *The Influence of Urbanization on Farm Ownership and Practice* (both, Wye College, Univ. of London: Dept. of Agricultural Economics, 1966), pp. 63 and 60 f., respectively.

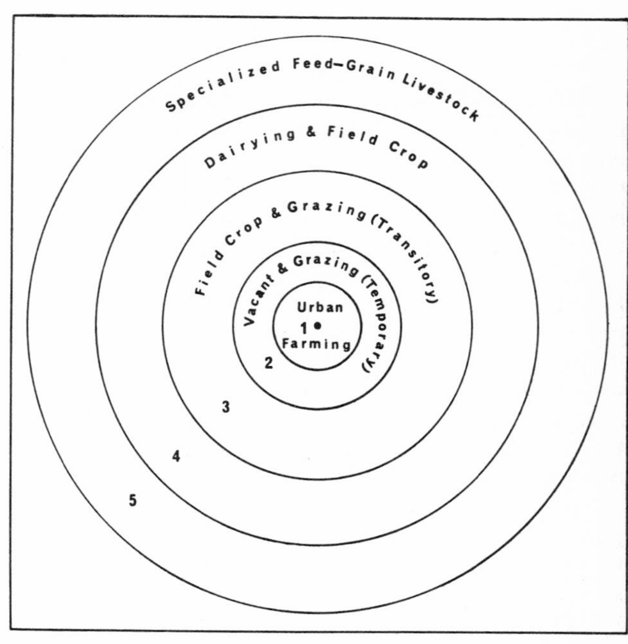

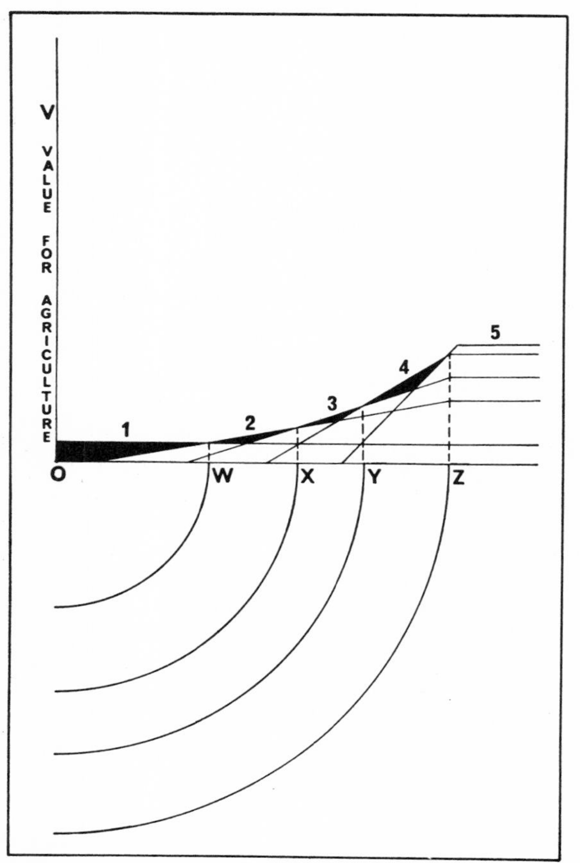

Figure 4. A (above): *A theoretical sequence of land uses around an expanding metropolitan area, according to R. Sinclair.*
B (opposite): *Relationship of value for agriculture and distance from urban area for competing land uses. As* O *(the urbanized area) is approached, the value of carrying on each of the particular types of agriculture decreases. The steepness of each* V*-slope is proportional to the intensity of the agricultural investment. Intercepts at* W, X, Y, *and* Z *indicate changes in land use. [By permission of the author; reproduced from the* Annals of the Association of American Geographers, *57 (1967), 80.]*

though its influence is not primarily that of a consuming market. The provision of services to food-processing industries and the supply of "day-haul" labor were two of the more important of these nonconsuming influences. Still later (1956) Phlipponneau commented on the centuries-long persistence of horticultural production around Paris. Other recent investigations have also brought out the important point that diminishing crop acreage on city margins can actually mask an intensification. Higbee has shown that fewer farmers have been able, through higher expenditures, shifts in production emphases, and more rationalized procedures, to increase production enormously in areas around cities in the Northeastern United States. No less impressive though on a smaller scale, the author has found, is the example of Los Angeles County in sextupling its farm production value from 1939 to 1959, despite losing a third of its acreage during the same period.[25]

Neither of these two blocks of opinion, however, precludes the possibility of a third view about the possibility of extensive agriculture on city margins: the development of an inner core of reverse intensity gradation within a larger Thünen pattern. A closer inspection of the reverse intensity pattern proposed by Sinclair (Fig. 4) reveals the beginning of a Thünen force at its margins, while an examination of the opposing literature shows that the existence of small and irregularly shaped plots of idle or only extensively used land on the margins of cities is often noted. Perhaps, then, the real problem is one of scale, and each side is basing its argument on an areal range that is ignored by the other.

Toward a General Theory of Location

The continuing debate over just how much evidence of Thünen-type zonation one can observe in the landscape and the general agreement that the pattern disintegrates as the economy of an area advances have inevitably stimulated speculation about a general theory of location, one that would take into account not only distance but all locational influences on the agricultural pattern. The most promising path for this research appears to be model building and testing. As observed earlier, the mathematical model is a tool that allows one to assess the effect of changes in any of the independent variables of a hypothesis, even if the relationships among the variables are extremely complex. Yet even such a model is challenged by the complexity of the relationships that must be explained by a theory of agricultural location.

Several attempts have been made at constructing a model that would account as much as possible for the spatial organization of agri-

[25] T. Brinkmann, *Die Ökonomik des landwirtschaftlichen Betriebes,* cited by L. Waibel, *Probleme der Landwirtschaftsgeographie* (Breslau: Hirt, 1933), p. 69; G. Pfeifer, *Die räumliche Gliederung der Landwirtschaft im nördlichen Kalifornien* (Leipzig: Hart, 1936), pp. 150 f.; M. Phlipponneau, *La vie rurale de la banlieue parisienne* (Paris: Colin, 1956); E. Higbee, "Megalopolitan Agriculture," in *Megalopolis,* ed. J. Gottmann (New York: Twentieth Century Fund, 1961), pp. 261–75; Gregor, "Urbanization of Southern California Agriculture," p. 274.

cultural activity.[26] All use the framework set up by Thünen as a basis for the model and then consider the importance of deviations from it. However, the question of how Thünen's model would look in the even more realistic conditions of changing technology and demand has still been little considered. Harvey raises the possibility of a dynamic Thünen model being applied to the advance of the agricultural frontier in newly developing countries, although he also admits that the development of any adequate growth model will be difficult because of the assumptions one is forced to posit about decision-making behavior. One of these is that complete information is available to the farmer, yet he must make his decisions in the face of unexpected weather conditions, sudden price changes, and other uncertainties. Game theory appears to offer at least a partial answer to the problem, for it enables the researcher to discover which of the alternatives open to the farmer in a decision-making problem he will rationally pick. Gould has effectively demonstrated this operation in his consideration of crop choices facing the Ghanaian farmer in an environment where the variability of precipitation is extreme.[27]

Game theory still cannot completely enlighten us on the decision-making process, however, for it operates on several assumptions that are untenable in an actual situation. The two most objectionable of these are that: (1) decisions are made solely on economic grounds and with a desire for optimum solutions; and (2) all existing information about market opportunities and technology is uniformly available and acceptable to the decision maker. One of the most vigorous critics of the first assumption has been Simon, who has proposed a model that replaces the goal of maximizing with the goal of "satisficing," of finding a course of action that is "good enough." To find this option, the decision maker first classifies the alternatives available to him on the basis of whether their outcomes will be satisfactory or unsatisfactory. Then, if the satisfactory outcomes can be ranked, the lowest satisfactory or "good enough" one will be chosen. The second assumption, that of uniformly available and acceptable information, has been countered in the model developed by Hägerstrand to describe the diffusion of agricultural innovations through an area in central Sweden. He shows that technical information may be taken up at different locations at different times.[28]

The growing interest of geographers and other spatially oriented scholars in behavioral models marks an important step toward a general theory of agricultural location. But it is still only a step. Most behavioral models have so far been constructed by sociologists and psychologists,

[26] A succinct summary of these attempts is provided by W. L. Garrison and D. F. Marble, "The Spatial Structure of Agricultural Activities," *Annals of the Association of American Geographers,* Vol. 47 (1957), 138 f.

[27] D. W. Harvey, "Theoretical Concepts and the Analysis of Agricultural Land-Use Patterns in Geography," *ibid.,* Vol. 46 (1966), 368; P. R. Gould, "Man Against His Environment: A Game Theoretic Framework," *ibid.,* Vol. 53 (1963), 291–94.

[28] H. A. Simon, *Models of Man* (New York: Wiley, 1957), pp. 204 f.; T. Hägerstrand, *Innovation Diffusion as a Spatial Process,* A. Pred, tr. (Chicago: Univ. of Chicago, Dept. of Geography, 1967).

who by and large have not specified their constructions to be pertinent to a spatial context. If once the combination of spatial and behavioral models is achieved, work will then have to be directed toward incorporating still another dimension, time. Whether these gaps can be bridged remains to be seen. Yet, in an age when interest in location theory is much greater than it was in the time of Thünen, the chances for success would seem to be at least better than they were then.

CHAPTER FIVE

The cultural impress

Immigrant Farmers

Farming is a way of life as well as an occupation, and this more cultural view has claimed a small but growing share of interest among researchers in agricultural geography. That an occupation represents the reaction of the individual to the demands of existence, but also that the type and success of this reaction is heavily influenced by such non-economic forces as race, nationality, religion, and psychology has long been acknowledged. For those interested in the geographic aspects, this fact has been particularly well typified by the fortunes of immigrant farmers. These groups have often brought sharp variations to the traditional ethnic and land use patterns of an area. Their impact has been especially noted in areas settled by European farmers, particularly in the United States and Canada, where the large majority went. Here the works of Kollmorgen stand out, particularly those on the "cultural islands" formed by immigrant farmers in the American South. The term "islands" is an apt one, for Kollmorgen discovered that these groups farmed quite differently from most farmers in the South. They worked harder, diversified their operations, kept at least some livestock, and avoided landlord-tenant and credit farming. They also built more and better farm buildings and made greater efforts to maintain them. Erosion in their fields was less evident. Yet all of this was accomplished on only the fair to poorer soils of the region. A wide variety of ethnic backgrounds was represented by these farmers, including Scandinavians, Dutch, Czechs, Poles, Hungarians, Italians, and especially Germans.[1]

[1] W. M. Kollmorgen, "A Reconnaissance of Some Cultural-Agricultural Islands in the South," *Economic Geography,* Vol. 17 (1941), 409–30; *idem,* "Agricultural-Cultural Islands in the South: Part II," *ibid.,* Vol. 19 (1943), 109–17.

Other writers have supplied further evidence for this thesis of ethnic superiority, in a variety of approaches and regional studies. Even before Kollmorgen had studied the South, Mackintosh had completed a detailed map of areas in the western Canadian plains occupied by foreign groups in 1929, which showed that the density of the rural population was more closely correlated with settlement by ethnic groups than with environmental differences. The highest densities he attributed to the larger families, smaller farms, and more intensive farming of the settlers from Central Europe. Some geographers have observed the singular success of certain groups in a particular farming industry, as Trewartha has done of the Swiss in the foreign cheese industry in Wisconsin and Perret of the Italians in dairying in California. Others have emphasized the importance of cultural attitudes toward farming in noting the decline of agricultural standards among European farmers when they were dispersed among other groups and not allowed to settle in ethnic units, as has Waibel in his research on the agricultural fortunes of certain foreign groups in Brazil. In contrast, settlement difficulties can sometimes challenge certain groups to make exceptional efforts, as Iwata has shown in his account of how the Japanese established themselves in California agriculture by cultivating lands that were considered so poor they were shunned by whites.[2]

Like Kollmorgen, many of these writers have particularly emphasized the vigor of the German farmer. Sauer earlier expressed a common opinion when he favorably compared the farmers of German stock in the United States with those of Anglo-Saxon origin. Both Sears and Cozzens, in their observations of German farmers in South Carolina and Missouri, found much less soil destruction on the German farms despite the general physical similarity of the study area. Farmers from Germany or of German ancestry have also been presented favorably in other parts of the world, as by James in his study of European colonies in southern Brazil and by Ehlers in his monograph on pioneer farming in the Peace River Valley of Canada.[3]

Yet, despite the many apparently close correlations between ethnic variations and farming abilities, a definitive conclusion is still far from being reached. The importance of traditional skills has been questioned

[2] W. A. Mackintosh, *Prairie Settlement, the Geographical Setting,* Vol. 1 of *Canadian Frontiers of Settlement,* ed. W. A. Mackintosh and W. L. G. Joerg (Toronto: MacMillan, 1934), pp. 81–84; G. T. Trewartha, "The Green County, Wisconsin, Foreign Cheese Industry," *Economic Geography,* Vol. 1 (1925), 296 f.; M. E. Perret, *Les colonies tessinoises en Californie* (Lausanne: Rouge, 1950); L. Waibel, "European Colonization in Southern Brazil," *Geographical Review,* Vol. 40 (1950), 545; M. Iwata, "The Japanese Immigrants in California Agriculture," *Agricultural History,* Vol. 36 (1962), 31, 37.

[3] C. Sauer, "The Settlement of the Humid East," in *Climate and Man,* ed. G. Hambidge (Washington, D.C.: U.S. Govt. Printing Office, 1941), p. 165; P. B. Sears, *Deserts on the March* (Norman: Univ. of Oklahoma, 1935), p. 58; A. G. Cozzens, "Conservation in German Settlements of the Missouri Ozarks," *Geographical Review,* Vol. 33 (1943), 286–98; P. E. James, "The Expanding Settlements of Southern Brazil," *ibid.,* Vol. 29 (1939), 601–26; E. Ehlers, *Das nördliche Peace River Country, Alberta, Kanada* (Tübingen: Univ. Tübingen, 1965), pp. 133–36, 179–82.

on several counts. Johnson points out that Germans who became farmers in the United States were not always peasants in Germany, nor did they start to farm immediately after arrival. Also, many farming specialties brought into the country by one group were quickly and successfully adopted by other groups. This is the conclusion one can draw from the findings of both Lewthwaite and Durand, who discovered that the presumed close correlation of national groups in the American Dairy Belt with manufacture of "national" cheese types was nonexistent and that a variety of national groups engaged in similar farming occupations. Furthermore, even those like Kollmorgen who have advanced the argument of the advantage of cultural inheritance have cautioned that immigrant farmers had to adjust at least partially to prevailing farming systems in their areas if they were to succeed.[4]

Pure chance seems to have played a good part in the apparent success of an ethnic group in a particular type of farming. The rapid rise of Americans of Japanese ancestry in the Hawaiian dairying industry is explained by Durand as a result of their recruitment to take the place of Portuguese who had left the farm for work at military installations, while the almost complete absence of farmers of Japanese extraction from California dairying is attributed by Gregor to earlier white prejudice and the later attraction of other and better-paying jobs. The effect of location on the fortunes of immigrant farm groups is mentioned by James and Augelli, who feel that differing degrees of access to areas with more opportunities for speculative farming or urban employment are the principal reasons why some farm settlements have failed and others succeeded, sometimes within the same ethnic group.[5]

The popular belief that German immigrants in particular usually turn out to be the best farmers has been critically viewed. Kollmorgen himself relates instances of German settlers doing no better in parts of the South than other groups, and Waibel contrasts the failure of Germans with the success of the Dutch in an area of southern Brazil. In the matter of soil conservation, Brown claims that German, as well as Polish, farmers in the Upsala community of Minnesota have been slower in adopting modern practices than their Scandinavian counterparts. Nor is the usual statement that Germans maintain neater farms and build higher-quality buildings, particularly barns, completely accepted. Durand found no correlation between the quality of barns and German settlement in southeastern Wisconsin; the differences that did exist were more

[4] H. B. Johnson, "The Location of German Immigrants in the Middle West," *Annals of the Association of American Geographers,* Vol. 41 (1951), 40; G. R. Lewthwaite, "Wisconsin Cheese and Farm Type: A Locational Hypothesis," *Economic Geography,* Vol. 40 (1964), 96 f.; L. Durand, Jr., "Italian Cheese Production in the American Dairy Region," *Annals of the Association of American Geographers,* Vol. 24 (1948), 217–30; W. M. Kollmorgen, "Immigrant Settlements in Southern Agriculture," *Agricultural History,* Vol. 19 (1945), 70.

[5] L. Durand, Jr., "The Dairy Industry of the Hawaiian Islands," *Economic Geography,* Vol. 35 (1959), 235; H. F. Gregor, "Industrialized Drylot Dairying: An Overview," *ibid.,* Vol. 39 (1963), 306; James, *op. cit.,* pp. 620 f.; J. P. Augelli, "Cultural and Economic Changes of Bastos, A Japanese Colony on Brazil's Paulista Frontier," *Annals of the Association of American Geographers,* Vol. 48 (1958), 19.

the result of individual differences among the original settlers of all groups or of the individual differences of their American descendants. The most vigorous attack on the belief in German farming superiority, however, has come from Lemon. After combing the historical literature on German, English, and Scotch-Irish farmers in Pennsylvania during the eighteenth century, he concludes that this belief derives from a stereotyped image of the German peasant held by Englishmen and perpetuated to the present day. Furthermore, he maintains, national character as a concept may have use in explaining architectural styles and food preparation, but it has little bearing on the comprehension of problems relating to agriculture.[6]

Still another basis for reservations about ethnic superiority arises from a consideration of the influences of variations in the quality of the environment. Although Kollmorgen's studies emphasize the cultural role, they also include several examples of significant contrasts in farming fortunes within the same ethnic group as a result of environmental differences. Undoubtedly, the tendency of many immigrant groups to settle in environments similar to their homelands has played a good part in their successes—illustrated by the Dutch reclamation of low-lying, swampy soils in Michigan and Ontario,[7] although such environmental choices may also be construed as a part of the ethnic inheritance. Environments are not always selected solely on the basis of agricultural desirability, however, and sometimes the other desired features lead to only meager farm returns. The most impressive example of such a mismatching in the United States is probably that made by Finnish farmers, the bulk of whom chose areas having forests, lakes, and a cool climate—which also had poor soils and a short growing season.[8]

How long immigrant farm groups can maintain themselves as an entity is still another unsettled point. Kollmorgen notes that some immigrant groups have existed in the South for more than a century. He feels that the present trend toward larger and more mechanized farms will only reinforce the numbers of newcomers to these groups because they have more of the needed capital than the native farmers. In contrast, one may infer from Durand's statements that earlier patterns of ethnic groupings in the American Dairy Belt, if not dissolved, are at least in the process of dissolution. He found that dairying was a spe-

[6] Kollmorgen, "A Reconnaissance of Some Cultural-Agricultural Islands," pp. 414, 419; Waibel, *op. cit.*, pp. 539 f.; R. H. Brown, "The Upsala, Minnesota, Community: A Case Study in Rural Dynamics," *Annals of the Association of American Geographers,* Vol. 57 (1967), 281; L. Durand, Jr., "Dairy Barns of Southeastern Wisconsin," *Economic Geography,* Vol. 19 (1943), 42–44; J. T. Lemon, "The Agricultural Practices of National Groups in Eighteenth-Century Southeastern Pennsylvania," *Geographical Review,* Vol. 56 (1966), 493–96.

[7] A. Sas, "Dutch Concentrations in Rural Southwestern Ontario During the Postwar Decade," *Annals of the Association of American Geographers,* Vol. 48 (1958), 192 f.; L. Durand, Jr., "The Lower Peninsula of Michigan and the Western Michigan Dairy Region: A Segment of the American Dairy Region," *Economic Geography,* Vol. 27 (1951), 176.

[8] E. Van Cleef, "The Finns of the Pacific Coast of the United States, and Consideration of the Problem of Scientific Land Settlement," *Annals of the Association of American Geographers,* Vol. 30 (1940), 25–38.

cialty of American farmers of practically all ancestries. Similar reports have come from other areas. Most of the researchers point out that the drying up of the immigrant flow to established colonies is reinforcing the drain to the cities and other places with more speculative opportunities. Perret notes the decline of Swiss dairymen in California soon after the passing of restrictive immigration legislation, and Augelli sees the decreasing immigration from Japan as one of the reasons why the absorption of Japanese colonies in Brazil appears to be but a matter of time. A better future is seen by Sas for Dutch farmers in southwestern Ontario, but only because both the Netherlands and Canada find it to their advantage to continue to promote the flow of migrants.[9] The disappearance of the immigrant group from the rural landscape, however, still would not necessarily mean an evaporation of its agricultural influence as well. The force of example and emulation, particularly when native farmers are in economic crisis and seeking solutions, has sometimes enabled immigrant farmers to project their system of farming well beyond their area of settlement.[10]

Religious Influences

Despite their close association with other cultural stimuli, religious motives in agricultural geography have received less attention than the other cultural motivations. Those who have approached the problem, principally anthropologists, sociologists, and cultural geographers, have focused more on non-Western agricultural forms, because here the influences of religion are much more direct. Nevertheless a wide variety of definite or possible religious influences has been considered. The research has produced little major debate as yet, probably because we are still in an early stage of investigation.

Relationships between farming operations and religious motives have been shown to be remarkably close in the simpler agricultural societies. Writers like Deffontaines, Malinowski, and Conklin have given impressive details of how religious rites and steps in the farming cycle are as one among groups in the lower latitudes. This symbiosis results in agricultural inefficiencies, for religious patterns which may have had a sound rationale at one time have a tendency to become fixed practice, no matter what changes occur in the physical or social environments. And, of course, not all such patterns were developed by correct reasoning in the first place. Even where religions are more sophisticated, they may actually create agricultural problems. Sopher observes that the Christian calendar noticeably retarded the introduction of new farming systems in Europe, and that the Islamic calendar, by ignoring the annual climatic and biological cycles, disrupts the farm calendar of agri-

[9] Kollmorgen, "Immigrant Settlements," pp. 77 f.; L. Durand, Jr., "The American Dairy Region," *Journal of Geography,* Vol. 48 (1949), 6; Perret, *op. cit.;* Augelli, "Cultural and Economic Changes," p. 19; Sas, *op. cit.*

[10] Kollmorgen, "Agricultural-Cultural Islands," pp. 112 f.; Cozzens, *op. cit.*, pp. 287, 291.

cultural folk like the fellahin of Egypt and the rice farmers of East Pakistan.[11]

Religion can also inspire a level of farming fully competitive with the most rationalized systems, by stressing the values of rural life. This is practically the unanimous opinion of those who have investigated the often remarkable showing of immigrant farm groups like the Old Order Amish and the Hutterites.[12] The deification of agriculture is also usually combined with a desire for isolation, so that the agricultural frontier is often advanced, not uncommonly into rigorous environments. Warkentin's documentation of the success of the Mennonites on the Canadian prairies is but one of several accounts of how religious zeal has expanded the ecumene.[13] Yet even among such inspired groups, religious tenets do not always favor the best farming operations. Thus, while the Hutterites favor extensive use of machinery in the large fields which result from their communal organization, the Old Order Amish have been faced with increasing schisms because its members find it difficult to observe the strictures on the use of machinery and yet compete with more mechanized farmers.[14]

The imprint of religion on field and settlement patterns has also been noted in several ways. Farm villages have been described as the most impressive visible evidence of the desire of religious adherents to maintain their solidarity, although religious motives have obviously not been the only or even the primary explanation for many group settlements. Nevertheless, the majority of group rural settlements have at least a partial religious base and their distribution is worldwide, with particularly important concentrations in Europe and Southeast Asia. None of this means that religion cannot encourage a dispersed or smaller rural population. Deffontaines claims that the wide scattering of temples for different gods in Tibet is a major cause of dispersion and nomadism, and Hart and Mather have suggested that the stern work demands of fundamentalist sects may well have reduced the need for as many farmers in

[11] P. Deffontaines, *Géographie et religions* (Paris: Gallimard, 1948), pp. 216–19; B. Malinowski, *Coral Gardens and Their Magic: A Study of the Methods of Tilling the Soil and of Agricultural Rites in the Trobriand Islands*, 2 vols. (London: Allen & Unwin, 1935); H. C. Conklin, *Hanunoo Agriculture; A Report on an Integral System of Shifting Cultivation in the Philippines* (Rome: Food and Agricultural Organization, 1957); D. E. Sopher, *Geography of Religions* (Englewood Cliffs, N.J.: Prentice-Hall, 1967), pp. 22, 39.

[12] The classic study of the Amish-farming symbiosis is W. M. Kollmorgen's *Culture of a Contemporary Rural Community: The Old Order Amish of Lancaster County, Pennsylvania*, Department of Agriculture, Bureau of Agricultural Economics, Rural Life Studies No. 4 (Washington, D.C.: U.S. Govt. Printing Office, 1942). M. P. Riley and J. R. Stewart are particularly enlightening in *The Hutterites: South Dakota's Communal Farmers*, Rural Sociology Dept., South Dakota State Univ. Bulletin No. 530 (Brookings, 1966).

[13] J. Warkentin, "Mennonite Agricultural Settlements of Southern Manitoba," *Geographical Review*, Vol. 49 (1959), 342–68.

[14] M. P. Riley, *The Hutterite Brethren: An Annotated Bibliography with Special Reference to South Dakota Hutterite Colonies*, Agricultural Experiment Station, South Dakota State Univ. Bulletin No. 529 (Brookings, 1965), pp. 18 f.; D. A. Price and L. F. Fleming, "The Old Order Amish Community of Arthur, Illinois," *Bulletin of the Illinois Geographical Society*, Vol. 7 (1965), 22.

one part of the American tobacco region as are needed in other parts.[15]

Farm fragmentation and smaller fields can also be traced in part to the influence of religion on village formation, although only the more immediate reason for the pattern, the problem of distance, is usually given. An impressive example of a direct influence of religion on field patterns is related by Sopher: Islam's rules of inheritance require land and other property to be divided equally among the sons, with a smaller share going to the daughters. The result has been excessive fragmentation and a corresponding decrease in farming efficiency. Others believe that many kinds of field borders in Europe can possibly be explained by old religious customs, which often forbade cultivation of the marginal strips as sacrilegious. Applied to wooded areas, such a proscription might well prove to have been originally responsible for the present hedgerows, and thus also contributory to the contrast between open and enclosed fields.[16]

There also seems little doubt that much of the plant and animal pattern of the world can be at least partly explained by religious beliefs. Haudricourt and Hédin would go so far as to ascribe a religious origin to all plants, since, according to them, all cultivated plants can be regarded "in a certain sense" to have been originally considered as sacred, as much for their actual or supposed properties as for their rarity or strangeness.[17] The ensuing movements of these plants have also in several cases been closely related to the migrations of peoples of similar beliefs, as was the spread of the vine over the Old World with Christian migrations and the extension of citrus over the Mediterranean area with Jewish migrations.[18] Food taboos vary from religion to religion, and so have attracted the attention of those who are interested in the reflections of these taboos in the agricultural landscape. Perhaps the most extensive study to date made of the subject has been that by Simoons, who explored the topic of taboos and other attitudes toward animal foods in the Old World. The absence of the pig in Islamic areas has been of special interest, and Fickeler uses the topic to deal directly with the knotty problem of how important religion is as a motivation in comparison with other stimuli. He argues for the primacy of the religious role over that of health because there are adherents of other religions in the subtropics and tropics who prize pork and yet have even stricter rules of ritual purity. To those who would try to delimit agricultural areas on the basis of taboos, however, Sopher offers a cautionary footnote, observing that their strength varies with the distance from the religious hearth.[19]

[15] Deffontaines, *op. cit.*, p. 130; J. F. Hart and E. C. Mather, "The Character of Tobacco Barns and Their Role in the Tobacco Economy of the United States," *Annals of the American Geographers*, Vol. 51 (1961), 291, fn. 24.

[16] Sopher, *op. cit.*, p. 44; Deffontaines, *op. cit.*, pp. 223 f.; L. Aufrère, "Systèmes agraires des Iles Brittaniques," *Annales de géographie*, Vol. 44 (1935), 388.

[17] A. G. Haudricourt and L. Hédin, *L'Homme et les plantes cultivées* (Paris: Gallinard, 1943), p. 90.

[18] Deffontaines, *op. cit.*, pp. 209–14; E. Isaac, "Influence of Religion on the Spread of Citrus," *Science*, Vol. 129 (1959), 179–86.

[19] F. J. Simoons, *Eat Not This Flesh* (Madison: Univ. of Wisconsin, 1961); P. Fickeler, "Fundamental Questions in the Geography of Religions," *Readings in Cultural Geography*, ed. P. L. Wagner and M. W. Mikesell (Chicago: Univ. of Chicago, 1962), p. 115; Sopher, *op. cit.*, p. 37.

Rural Social Structure

The stratification of rural society has also received comparatively less attention in agricultural geography, the research on the topic being more impressive for its range than its depth.

Farm size has been seized upon by most investigators as the best over-all indicator of social layering. Pfeifer and other German scholars have used it as a guide to their regionalization schemes of the rural society of Germany. Coquery and Guérémy have also found a close correlation between the rural social types and property sizes of the Paris Basin. One may also infer such an interrelationship from the account by Prunty of a social hierarchy among share and tenant farmers in the lower Mississippi Valley, based upon use or ownership of mechanized farm equipment. Similarly, Schul, in his Philippine study of the top social layer on the plantations, the planters, found social overtones in a hierarchy based partly on plantation size.[20]

Correlations between social groups and types of land use have also been made. In the heart of the Flint Hills–Bluestem pasture lands of Kansas, Kollmorgen and Simonett have mapped a close correspondence of large, absentee urban owners with ranching areas and small, residential rural owners with most of the croplands. Differences between the land uses of large and small landholders have also been reported within areas of intensive agriculture. In the Avon Valley of the English Midlands, small growers raise a much larger variety of truck crops than the large operators, according to Buchanan; and in Malaysia, Gosling reaffirms the relationship in his findings that smaller farmers devote more land to those subsistence crops, such as fruits and vegetables, that they can sometimes sell in the markets.[21]

In many areas of intensive agriculture, however, differences in the size of landholdings are insignificant and there is still an intricate social structure. Other indicators have to be found. Hornberger, in his study of the Georgenberg area of southwestern Germany, made a detailed land use map (1:3600) and compared it with the distribution of occupations provided by land registry records as well as with the areal variation in

[20] G. Pfeifer, "The Quality of Peasant Living in Central Europe," in *Man's Role in Changing the Face of the Earth,* ed. W. L. Thomas, Jr. (Chicago: Univ. of Chicago, 1956), pp. 246–49; M. Coquery, "La structure de la propriété foncière dans la plaine de France et son évolution," in *Etudes de géographie rurale,* ed. P. George (Saint Cloud: Société Amicales des Anciens Elèves de l'E.N.S. de Saint-Cloud, 1959), pp. 35–53; P. Guérémy, "Inégale concentration de la propriété et de l'exploitation dans le Hurepoix," *ibid.,* pp. 109–35; M. C. Prunty, Jr., "Soybeans in the Lower Mississippi Valley," *Economic Geography,* Vol. 27 (1951), 310 f.; N. W. Schul, "A Philippine Sugar Cane Plantation: Land Tenure and Sugar Cane Production," *ibid.,* Vol. 45 (1967), 157–69.

[21] W. M. Kollmorgen and D. S. Simonett, "Grazing Operations in the Flint Hills–Bluestem Pastures of Chase County, Kansas," *Annals of the Association of American Geographers,* Vol. 55 (1965), 280 f.; K. Buchanan, "Modern Farming in the Vale of Evesham," *Economic Geography,* Vol. 24 (1948), 239; L. A. P. Gosling, "The Relationships of Land Rental Systems to Land Use in Malaya," *Papers of the Michigan Academy of Arts and Letters,* Vol. 44 (1958), 330.

production orientation, whether for the home or the market. His conclusions were that in one section of the area renters and part-time farmers concentrated more on intensive crops than did either the large or small full-time farmers, but that in the remaining portion of the study area cultivators in general farmed even more intensively and sent more of their crops to market. This contrast Hornberger explained as the result of less social differentiation and the lack of a more tradition-bound farm group in the latter area.[22]

Geographers have not remained aloof from the problems engendered by extreme economic contrasts between the several social groups of an area, particularly where large and small landholdings are closely juxtaposed. Already in 1924, Colby was noting that irrigation in parts of the San Joaquin Valley of California had been rebuffed by the larger farmers because they wished to continue livestock and grain farming. The use of a plow that predisposes the field to soil blowing and the practice of monoculture instead of diversification are listed by Kollmorgen and Jenks as two of the reasons for the antagonism of the resident wheat farmer in the Great Plains toward the absentee, or "suitcase," farmer. Kollmorgen also stresses the success in the American South of farming groups with a homogeneous rural structure in contrast to the fortunes of those who are in a clearly structured system, one in which no "agricultural ladder" exists and the top stratum has little desire to work in the fields. Farm consolidation is hampered by the differences in demographic rates between large and small farmers in French Normandy, according to Frémont. Since the resulting labor surplus ensures takers, the large owners can exact high rates of rent and thus make as much or more money on several tenant-operated farms as they could by operating a single large and more efficient farm. And in Dominica, an island in the West Indies, Finkel observes, the absence of a land tax encourages the large owners to hold idle land for an indefinite period, whereas the small farmers are forced to move onto lands that are much too prone to erosion.[23]

What appears to be the biggest problem, however, is the increasing economic advantage of large-scale farming at the very time that the demands for the breakup of large properties are at their strongest. Chardon and Augelli graphically describe the dilemma in their articles on plantations, although Gregor sees hope in a future plantation with per-

[22] Th. Hornberger, "Anbaugefüge und Sozialstruktur am Georgenberg südlich Reutlingen," in *Atlas der deutschen Agrarlandschaft,* ed. E. Otremba (Wiesbaden: Steiner, 1965), 2d Lfg., Teil IV, Blatt 10.

[23] C. C. Colby, "The California Raisin Industry: A Study in Geographic Interpretation," *Annals of the Association of American Geographers,* Vol. 14 (1924), 74 f.; W. M. Kollmorgen and G. F. Jenks, "Suitcase Farming in Sully County, South Dakota," *ibid.,* Vol. 48 (1958), 36 f., 40; Kollmorgen, "A Reconnaissance of Some Cultural-Agricultural Islands," pp. 412, 424; *idem,* "Agricultural-Cultural Islands," pp. 116 f.; A. Frémont, "Pays de Coux et bordure meridionale du pays d'Ouche," in *Etudes de géographie rurale,* p. 100 f.; H. J. Finkel, "Patterns of Land Tenure in the Leeward and Windward Islands and Their Relevance to Problems of Agricultural Development in the West Indies," *Economic Geography,* Vol. 40 (1964), 172.

sonnel that is more industrially oriented in its values and a policy that is more responsive to human needs. For farmers in general, George and Dumont see some form of cooperative organization, combined with land reform, as the answer, George favoring the example of the Soviet state farms and Dumont that of the Israeli farms.[24]

The structure of rural society is also usually translated into the geographic structure of rural settlements. India, with its abundance of farm villages and rigid caste compartmentalization, offers an especially attractive opportunity for investigating these societal-morphological relationships. Excellent evidence of these relationships is to be found in Spate and Deshpande's study of a large Indian village, Aminbhavi. Here the landlords—and the leading castes—live in large compounds and on the best sites in the center of the community. The tenants and landless agricultural laborers take up the remainder of the settlement area, although even they are separated into their own districts, differentiated by both quality and style of house. Because agricultural castes have no enviable position in Indian village society, many farmers are not reluctant to leave the main site and settle near their plots, thus accentuating their eccentric location and contribution to a multiplication of outlying hamlets. The same outward gradation in social and economic levels has been reported by Yang in China, Hall in Japan, and Otremba in Germany.[25] Another type of rural settlement structure, but one where the distribution of houses is strictly based on familial lineage, has frequently been observed in tribal areas of Latin America, Africa, and Southeast Asia.[26] Reflections of social stratification in settlements associated with plantations and closely related large farms of an industrial character have been noted by many geographers.[27] But although these and other studies commonly illustrate with maps and photographs the various arrangements of the buildings housing married and single workers, technicians, administrators, and foremen or owner, few have analyzed the patterns in any depth. A successor to the kind of investigation carried out by Platt (1943) in his extensive study of individual farm types and associated settlement structures in Latin America still awaits the pen.[28]

[24] R. E. Chardon, "Hacienda and Ejido in Yucatan: The Example of Santa Ana Cucá," *Annals of the Association of American Geographers,* Vol. 53 (1963), 174–93; J. P. Augelli, "Patterns and Problems of Land Tenure in the Lesser Antilles: Antigua, B.W.I.," *Economic Geography,* Vol. 29 (1953), 262–67; H. F. Gregor, "The Changing Plantation," *Annals of the Association of American Geographers,* Vol. 55 (1965), 228–35; P. George, *Précis de géographie rurale* (Paris: Presses univ. de France, 1963), pp. 126–34; R. Dumont, *Voyages en France d'un agronome,* 2d ed. (Paris: Médicis, 1956), pp. 218–20.

[25] O. H. K. Spate and C. H. Deshpande, "The Indian Village," *Geography,* Vol. 37 (1952), 142–52; M. C. Yang, *A Chinese Village: Taitou, Shantung Province* (New York: Columbia Univ., 1950); R. B. Hall, "Some Rural Settlement Forms in Japan," *Geographical Review,* Vol. 21 (1931), pp. 113 f.; E. Otremba, *Die deutsche Agrarlandschaft* (Wiesbaden: Steiner, 1961), pp. 41 f.

[26] G. Schwarz, *Allgemeine Siedlungsgeographie* (Berlin: de Gruyter, 1961), pp. 83–89.

[27] Particularly M. Prunty, Jr., "The Renaissance of the Southern Plantation," *Geographical Review,* Vol. 45 (1955), 484 f.

[28] R. S. Platt, *Latin America: Countrysides and United Regions* (New York: McGraw-Hill, 1943).

Rural Man

Intensive inquiry into the geographic nature of rural society on the basis of psychological rather than physiognomic evidence has been undertaken even less. Students most interested in the psychological aspects of the farmer, particularly the rural sociologists, have concerned themselves, more often than not, with the traits of a generalized "rural man," and references to areal differences in the psychology of rural groups have usually been confined to brief comparisons of commercial and subsistence farmers. Where observations and analyses have been more intensive, they have been limited to a particular group, as in Luebke's and Hart's account of southern Appalachian farmers and their rural ethos, or to several groups in the context of a particular psychological situation, as in Bowman's comprehensive review of farmers' attitudes toward frontier farming.[29]

The gradual increase of these more local, but also more intensive, studies, however, has also encouraged some researchers to feel that the time has come for at least some tentative consideration of worldwide patterns in rural psychology. Fully admitting the difficulties of generalizing about the many forces that have shaped the outlook of rural groups, Otremba nevertheless selects three criteria which he feels can eliminate the side effects of race, religion, and nationality and thus provide a basis for a regionalization based solely upon psychological characteristics. These indices are land ownership status and attitudes toward land and money. Northwestern Europe is designated as a region of farmers who generally own their own land and, although commercially oriented, still carry on some subsistence agriculture. A strong attachment to their land and to the traditional way of doing things, irrespective of economic considerations, is characteristic. Toward the Mediterranean, land hunger among the tradition-conscious peasants increases as rural society becomes increasingly feudal, and landlords with a curious blend of capitalistic motives and the traditional land roots of the peasant become more numerous. Toward the east, the small farmer gradually gives way to a special type of individual in which the characteristics of the industrial worker mix with those of the peasant.

Anglo-America, Australia, and that part of South Africa settled by English settlers contains farmers who are rationally and commercially oriented to the extreme; farming is looked upon primarily as an occupation rather than a way of life. Although land ownership is widespread, sentimental attachment to the land is not nearly so intense as it is in Europe and other areas where a long farming tradition exists. In contrast, farmers in populous Southeastern Asia have long prized the land as an objective in itself, although the great majority have been only tenants. Human labor reaches its peak of intensity in this area, but in an economy

[29] B. H. Luebke and J. F. Hart, "Migration From a Southern Appalachian Community," *Land Economics,* Vol. 34 (1958), 44–53; I. Bowman, *The Pioneer Fringe* (New York: American Geographical Society, 1931).

that often contrasts with the rational and capitalistic outlook prevalent in much of Europe and North America. Yet changes are multiplying in this area, the most drastic being the rise of the Chinese commune, a form of social organization that goes well beyond the Soviet collective farm. Latin America, much of Africa, and Indonesia have two major rural groups, those who work their own lands and those who labor on plantations. The indigenous farmers commonly view their activities as the only way of life, one that is little influenced by considerations of economics or rationality. The primary concern of the plantation workers, however, is the need for land and an improvement of their economic condition. Yet changes are going on in these areas also: governments are introducing land reforms, often in the form of cooperative or communistic farms.

These characterizations are only broad sketches of a much more complex framework of rural psychological regions, as Otremba himself admits and abundantly illustrates. More localized and detailed regionalization will obviously have to be attempted. The system of rural culture regions set up by Taylor, Raper, and McKain for the United States is evidence that this need is now well recognized.[30]

[30] E. Otremba, *Allgemeine Agrar- und Industriegeographie* (Stuttgart: Franckh, 1953), pp. 98–120; C. C. Taylor *et al., Rural Life in the United States* (New York: Knopf, 1949), pp. 329–491.

CHAPTER SIX

Political reflections

Land Use Controls

Remarkably little research has been done on the effects of political decisions on the rural landscape, compared with work on other aspects of agricultural geography; but, as in the study of cultural influences, a variety of topics has nevertheless been considered, many promising objectives having already been outlined by Whittlesey in the mid-Thirties.[1] Although depth has been the principal lack, the deficiency is now being repaired, particularly in the study of governmental controls and their effects on individual crop and livestock patterns. This relationship has been studied on both the international and intranational, or regional, levels.

On the international level, the longest period of attention has been accorded the connections between colonial policies and the distribution of tropical crops grown on plantations. By 1937, Waibel had written an extensive monograph on such crops in Africa, showing how both the kind of crop and its areal extent were closely related to the area of political influence and the vigor and goals of colonial policy. Several years later he again dwelt on this relationship, maintaining that the division of West Africa into French and British spheres was dictated by the needs of the two countries for different sources of fat, France's for peanuts and Britain's for palm kernels. Most recently, Courtenay has correlated the aggressive economic policies of the British and the Dutch in Southeastern

[1] D. Whittlesey, "The Impress of Effective Central Authority upon the Landscape," *Annals of the Association of American Geographers,* Vol. 25 (1935), 85–97.

Asia with the rapid expansion of tea and rubber in their former territories.[2]

The political network of Europe and its pertinence to the agricultural pattern there has also interested researchers to some degree. Grotewold and Sublett provide an illuminating view of the many contrasts between the crop patterns of England and Germany which have resulted from Germany's greater restrictions of imports. These differences should not lead one to assume, however, that the English agricultural landscape has been only mildly influenced by national policies. Stamp relates the significant increase in improved land since the early Thirties to the increased governmental emphasis on self-sufficiency. Geographers interested in the European situation have also inquired into the effects of political influence on those characteristics of the agricultural pattern that are more functional than formal. Otremba has considered the relationship of Germany to its neighbors in the context of types of agricultural economies. Others, like Van Valkenburg, have emphasized the effect of supranational organizations on the international competition of agricultural products, while Enyedi has concerned himself with variations in national farming policies and their effects on the production levels of Eastern European countries.[3]

A variety of approaches to political-agricultural relationships is to be found in studies of other parts of the world, as well as Europe. The highly disruptive effect of the Indian-Pakistani partition on agriculture has stimulated studies of individual crop patterns, over-all agricultural economy, and irrigation in various affected areas.[4] Even among the normally genial relationships between the United States and Canada, geographers have not failed to find political antagonisms conducive to differences in farming systems, in the relationships between similar economies (dairying), in marketing procedures for the same products, and in rural settlement characteristics.[5]

[2] L. Waibel, *Die Rohstoffsgebiete des tropischen Afrikas* (Leipzig: Bibliographisches Institut, 1937); *idem*, "The Political Significance of Tropical Vegetable Fats for the Industrial Countries of Europe," *Annals of the Association of American Geographers*, Vol. 33 (1943), 120; P. P. Courtenay, *Plantation Agriculture* (New York: Praeger, 1965), pp. 144–203.

[3] A. Grotewold and M. D. Sublett, "The Effect of Import Restrictions on Land Use: The United Kingdom Compared with West Germany," *Economic Geography*, Vol. 45 (1967), 64–70; L. D. Stamp, "Nationalism and Land Utilization in Britain," *Geographical Review*, Vol. 27 (1937), 1–18; E. Otremba, *Die deutsche Agrarlandschaft* (Wiesbaden: Steiner, 1961), pp. 11–16; S. Van Valkenburg, "Land Use within the European Common Market," *Economic Geography*, Vol. 35 (1959), 1–24; G. Enyedi, "The Changing Face of Agriculture in Eastern Europe," *Geographical Review*, Vol. 57 (1967), 358–72.

[4] One of the most current and detailed treatments is A. A. Michel's *The Indus Rivers: A Study of the Effects of Partition* (New Haven: Yale Univ., 1967).

[5] Respectively: E. Dix, "United States Influences on the Agriculture of Prince Edward County, Ontario," *Economic Geography*, Vol. 26 (1950), 179–82; L. Durand, Jr., "The Lower Peninsula of Michigan and the Western Michigan Dairy Region: A Segment of the American Dairy Region," *ibid.*, Vol. 27 (1951), 166 f.; J. W. Watson, *North America: Its Countries and Regions* (London: Longmans, Green, 1963), 214–16.

Research on the influence of governmental decisions on the agricultural geography of an area has also focused on the resulting variations within countries. Regional studies of this kind have concentrated particularly on the United States, where serious problems of surplus production and extensive soil degradation have led to major governmental efforts at land use control. The effect has been a massive reduction in acreages of major crops. Erickson and Prunty have described in detail how a program for reducing cotton surpluses has fragmented the once extensive Cotton Belt. Similiarly, Hewes has shown that extensive reductions in acreages of wheat and other grains took place on the Central Great Plains between 1956 and 1960 as growers accepted a government offer to pay "rent" for land if it was withdrawn from cultivation and put to grass.[6]

Equally well documented are other changes in the landscape pattern resulting from government programs. The reduction of acreage has been accompanied by increasing concentration in the most favorable areas, so that losses in acreage have been effectively countered by increases in acre-yields. New land uses have been introduced into the abandoned areas, sometimes to a point where the regional economy has been completely transformed: notable examples are the increase of livestock raising in the Southeast and the spread of soybeans into the lower Mississippi Valley.[7] In still other areas, specialized crop patterns have been fossilized by strict acreage controls. Newcomers can obtain an acreage allotment only at the expense of established growers, and for this reason almost all applications are denied. Political restriction of the widely adaptable tobacco plant in the Southeast has been labeled "political determinism," and acreage restrictions on cotton based on "acreage histories" have been called a barrier to an even greater shift of cotton from the Southeast to the Southwest, where the crop can be produced in greater quantities and more efficiently.[8] Still, in both these cases the restriction does not appear absolute. Durand and Bird tell us that tobacco farms have increased in Tennessee because farmers are allowed to start raising tobacco if they have had experience in the practice during at least two of the preceding five years and are willing to pay a penalty for planting the first year. Hart believed that the increasingly heavier yields on Southwestern cotton farms would force the gov-

[6] F. C. Erickson, "The Broken Cotton Belt," *Economic Geography,* Vol. 24 (1948), 263–68; M. Prunty, Jr., "Recent Quantitative Changes in the Cotton Regions of the Southeastern States," *ibid.,* Vol. 27 (1951), 189–208; L. Hewes, "The Conservation Reserve of the American Soil Bank as an Indicator of Regions of Maladjustment in Agriculture, with Particular Reference to the Great Plains," *Wiener Geographische Schriften* (Festschrift Leopold G. Scheidl zum 60. Geburtstag, eds. H. Baumgartner *et al.,* Part 2), Nos. 24–29 (1967), pp. 331–46.

[7] M. Prunty, Jr., "Land Occupance in the Southeast: Landmarks and Forecast," *Geographical Review,* Vol. 42 (1952), 453–60; *idem,* "Soybeans in the Lower Mississippi Valley," *Economic Geography,* Vol. 27 (1951), 301–14.

[8] Prunty, *op. cit.,* 1952, p. 450; H. F. Gregor, "The Regional Primacy of San Joaquin Valley Agricultural Production," *Journal of Geography,* Vol. 61 (1962), 396 f.; D. C. Large, "Cotton in the San Joaquin Valley: A Study of Government in Agriculture," *Geographical Review,* Vol. 47 (1957), 365.

rnment to impose ever greater acreage restrictions, thus eliminating the mall cotton farmer who is more numerous in the Southeast, and Higbee nds that the process is well on its way toward completion.[9]

Although there are numerous references to the effect of federal ontrol on American agricultural patterns, there is a paucity of writings n the influence exerted by state and municipal agencies. There is little esides the articles of G. J. Fielding, in which he shows that state marketng laws and local zoning ordinances have preserved the Los Angeles nilkshed despite growing diseconomies.[10]

There are few studies of political influence on regional agricultural atterns outside the United States, but there is still a surprising range of indings. One of the more interesting is Boal and McAodha's account of he elimination, by government subsidy, of the influence of transport ost on the location and nature of milk production in northern Ireland. The freezing of land use patterns is again reflected in the areas of comparatively stable sheep population in Britain, which, according to Hart, orrelate closely with areas of heavy subsidization. Relations between he two portions of Pakistan have been strained by governmental pricng and by policies governing the export of the two competitive cash groups of the country—cotton in West Pakistan, and jute in East Pakistan. Graziers in the southern part of the Brazilian state of Bahia clear the orest and ignore the savannas, and owners of plantations do little to expand their holdings, all because, as Monbeig tells us, the financing olicy of the Bank of Brazil is more generous to the graziers. And among he various Spanish provinces, the effects of national agricultural policies ave been strong enough to convince Hofmeister that governmental acion will eventually resolve the traditional dilemma posed to researchers of discrepancies between conceptual and administrative regions.[11]

The farm unit, too, has received attention in the search for signifiant landscape reflections of governmental action, and recent and deailed studies of this kind bulk large in the total literature. This is most evident in articles analyzing the particulars and progress of political efforts at enlarging farms through consolidation so as to improve efficiency. Researchers have concentrated largely on Europe, where con-

[9] L. Durand, Jr., and E. T. Bird, "The Burley Tobacco Region of the Mounain South," *Economic Geography,* Vol. 26 (1950), 287; J. F. Hart, "Cotton Goes West in the American South," *Geography,* Vol. 44 (1959), 45; E. Higbee, *Farms and Farmers in an Urban Age* (New York: Twentieth Century Fund, 1963), 14 f.

[10] His most incisive work is "The Los Angeles Milkshed: A Study of the Poitical Factor in Agriculture," *Geographical Review,* Vol. 54 (1964), 1–12.

[11] F. W. Boal and B. S. McAodha, "The Milk Industry in Northern Ireland," *Economic Geography,* Vol. 37 (1961), 170–80; J. F. Hart, "The Changing Distribuion of Sheep in Britain," *ibid.,* 32 (1956), 273; G. V. Stephenson, "Pakistan: Discontiguity and the Majority Problem," *Geographical Review,* Vol. 58 (1968), 203 f.; P. Monbeig, "Les capitaux et la géographie," *Géographie Générale* (*Encyclopédie de la Pléiade,* Vol. 20), ed. A. Journaux *et al.* (Paris: Gallimard, 1966), p. 1518; B. Hofmeister, "The Impact of State Regulation on Spain's Major Olive-Growing Regions," in *Agricultural Geography: I.G.U. Symposium,* ed. E. S. Simpson (Liverpool: Department of Geography, Univ. of Liverpool, 1965), p. 21.

solidation has been most active, and studies include both national and regional views of the programs.[12]

The discrepancies between farm size prescribed by law and that dictated by environmental conditions have also been favorite themes of researchers. Americans interested in the problem of insufficient farm size still have as their benchmark the classic work by the historian Webb, *The Great Plains*, in which the long history of inadequate farms in that region is traced to laws enacted by legislators having no familiarity with the problem of aridity. Ehlers, in his study of past Canadian Prairie settlement, echoes this theme, as does Chisholm in his account of how Piedmontese law, suited to the more humid climate of northern Italy was also applied to the rest, and drier, part of the country. Other writers, such as Chardon, touch upon a related topic in their accounts of how the subdivision of large properties by the Mexican Government has created large numbers of farms that are too small for efficient operation.[13] A major reason for this misfortune is that most Mexicans live in only a very few parts of the country.

Colonization and Reclamation Projects

Agricultural colonization and land reclamation have been traditionally associated with governmental activity, an association which becomes more intimate as farmland grows scarcer and project costs increase. Research interest has correspondingly increased, and a respectable number of writings on the role of government along the agricultural fringe now exists. It is probably no surprise that scholars have paid most attention to government projects designed to reclaim lands from the desert or the sea, for here the results have been the most striking. The emphasis in studies of the dry lands has usually been on the many and often intricate patterns of water transferral and the nature of the hydraulic installations. The Western United States and West Pakistan have received some of the best attention of this kind.[14] A similar emphasis on the engineering efforts of governments marks many of the studies of projects for reclaiming wet or inundated lands, such on those on the reclamation and settlement of the Guadalquivir Delta of Spain and the coastal marshes

[12] Representative articles are J. Naylon, "Land Consolidation in Spain," *Annals of the Association of American Geographers*, Vol. 49 (1959), 316–73; and A. M. Lambert, "Farm Consolidation and Improvement in the Netherlands: An example from the Land Maas en Waal," *Economic Geography*, Vol. 37 (1961), 115–23.

[13] W. P. Webb, *The Great Plains* (Boston: Ginn, 1931; New York: Grosset and Dunlap, Grosset's Univ. Library reprint), pp. 385–452; E. Ehlers, "Landpolitik und Landpotential in den nördlichen Kanadischen Prärieprovinzen," *Zeitschrift fur ausländische Landwirtschaft*, Vol. 50 (1966), 44; M. Chisholm, *Geography and Economics* (London: Bell, 1966), pp. 214 f.; R. E. Chardon, "Hacienda and Ejido in Yucatan: The Example of Santa Anna Cucá," *Annals of the Association of American Geographers*," Vol. 53 (1963), 174–93.

[14] J. M. Roy, "Les grands plans d'aménagement hydraulique et la mise en valeur de l'ouest américain," *Revue canadienne de géographie*, Vol. 10 (1956), Part 1, 175–90; Vol. 11 (1957), Part 2, 3–26, and Part 3, 95–107; L. H. Gulick, Jr., "Irrigation Systems of the Former Sind Province, West Pakistan," *Geographical Review*, Vol. 53 (1963), 79–99.

of Northwestern Europe.[15] Where the reflection of governmental activity in the landscape has not been so stark, research attention has been more equally divided between the technological and administrative activities of government. The articles by Vanderhill on the efforts of Canadian provincial authorities to sponsor settlement of the northern agricultural fringe of the Prairies, particularly in Manitoba, show this more even blend particularly well.[16]

The critical contribution of governmental enthusiasm to successful colonization has also been a prime topic of interest. Some geographers have detailed the importance of a vigorous settlement policy in terms of the particular accomplishments of a project, as has Augelli for the colonization program in the Dominican Republic. Others have shown that differences in the political problems facing different nations can lead to marked contrasts in the intensity of politically supported programs. Eidt points out that boundary controversies have encouraged Peru to develop its eastern area more than Colombia, despite the fact that the eastern regions of the two countries are physically similar; McDermott believes that emigration helped to stimulate the Quebec Government to actively advance the farming frontier in the eastern part of the Ontario Clay Belt, while the same frontier in the western part retreats under the laissez-faire policy of the Ontario Government. Yet governmental encouragement may not always guarantee an advance of the agricultural frontier, as Ehlers observes in his comparison of recent settlements in Western Canada. The settlement programs in Manitoba and Saskatchewan have not been able to advance the frontier as fast as it has moved in Alberta and British Columbia, where governmental policies are less hospitable but where more arable land is available.[17]

Government land surveys have also left an impressive imprint on the agricultural landscape. This is clearest in the more recently settled lands, where surveys have usually preceded extensive settlement, and where occupation has not been prolonged enough to obscure the schematic pattern. In this respect, Canada and the United States have been particularly attractive areas of study. However, although geographers, agriculturists, sociologists, engineers, and other specialists have written on the subject, the total amount of literature is still surprisingly small considering the impact of the survey systems on current land use and settle-

[15] As by P. M. Enggass, "Land Reclamation and Resettlement in the Guadalquivir Delta–Las Marismas, *Economic Geography,* Vol. 44 (1968), 125–43; and P. Wagret, *Polderlands,* M. Sparks, tr. (London: Methuen, 1968).

[16] B. G. Vanderhill, "Post-War Agricultural Settlement in Manitoba," *Economic Geography,* Vol. 35 (1959), 259–68; *idem,* "The Success of Government-Sponsored Settlement in Manitoba," *Journal of Geography,* Vol. 61 (1962), 152–62.

[17] J. P. Augelli, "Agricultural Colonization in the Dominican Republic," *Economic Geography,* Vol. 37 (1961), 15–27; R. C. Eidt, "Comparative Problems and Techniques in Tropical and Semitropical Pioneer Settlement: Colombia, Peru and Argentina," *Yearbook of the Association of Pacific Coast Geographers,* Vol. 26 (1964), 38 f.; G. L. McDermott, "Frontiers of Settlement in the Great Clay Belt, Ontario and Quebec," *Annals of the Association of American Geographers,* Vol. 51 (1961), 261–73; E. Ehlers, *Das nördliche Peace River Country: Alberta, Kanada* (Tübingen: Geographisches Institut, Univ. Tübingen, 1965), 214–15; *idem,* "Landpolitik und Landpotential," p. 54.

ment patterns. What has been done at least provides a basic structure on which to build future research. The development of the North American rectangular system in its hearth areas of Ontario and the Middle West has been analyzed in detail by Schott, Taylor, and Pattison. It is also these three scholars, as well as Marschner and Hart, who have shown that the rectangular system in North America, contrary to common belief, has an unusual variety of orientations and outline. Deffontaines, Bartz, and Harris, among others, have provided much the same kinds of information for the long-lot system in North America, most widespread in Quebec.[18]

Neither research group has fully settled the problem of the origins of the two systems, although inheritance from France seems to be the favorite explanation for the long-lot pattern and the example of the Roman *centuriae* the most preferred argument for the rectangular pattern. Debate on the comparative merits of the two systems has so far favored the long-lot grid. The principal objections to the rectangular system have been its disregard of variations in the terrain and its furtherance of a settlement pattern that greatly increases the cost of services and heightens isolation. Thrower has found, in his comparison of systematically and unsystematically surveyed lands in Ohio, that the rectangular system has encouraged more farm fragmentation through its finer net of roads.[19]

Studies of schematic land survey patterns on other continents have uncovered numerous examples of both long-lot and rectangular designs, although many are not the results of state-planned settlement projects and many are only small interruptions in an intricate and irregular network of field patterns. This is commonly the case in the long-settled and intensively farmed regions of Europe and Southeast Asia.[20] Geometric survey patterns have been found to be far more evident in widely dispersed but more recently settled areas like Hokkaido, the Argentine Pampa, and Australia. Yet some of the latest writings again emphasize the limits of generalization: Meinig recalls the experience of Schott in southern Ontario in his discovery of a rectangular system in the state of South Australia, which is considerably more varied than the standard pattern;

[18] C. Schott, *Landnahme und Kolonisation in Canada am Beispiel Südontarios* (Kiel: Geographisches Institut, Univ. Kiel, 1936), pp. 83–109; G. Taylor, "Towns and Townships in Southern Ontario," *Economic Geography*, Vol. 21 (1945), 88–103; W. D. Pattison, *Beginnings of the American Rectangular Land Survey System* (Chicago: Dept. of Geography, Univ. of Chicago, 1957); F. J. Marschner, *Land Use and Its Patterns in the United States*, Dept. of Agriculture Handbook No. 153 (Washington, D.C.: U.S. Govt. Printing Office, 1959); J. F. Hart, "Field Patterns in Indiana," *Geographical Review*, Vol. 58 (1968), 450–71; P. Deffontaines, "Le rang, type de peuplement rural du Canada français," *Cahiers de Géographie*, Vol. 5 (1953), 3–32; F. Bartz, "Französiche Einflüsse im Bilde der Kulturlandschaft Nordamerikas," *Erdkunde*, Vol. 9 (1959), 286–305; R. C. Harris, *The Seigneurial System in Early Canada* (Madison: Univ. of Wisconsin, 1966), pp. 117–38.

[19] C. P. Barnes, "Economies of the Long-Lot Farm," *Geographical Review*, Vol. 25 (1935), 298–301; T. L. Smith, *The Sociology of Rural Life*, 3d ed. (New York: Harper, 1953), pp. 263–73; N. J. W. Thrower, *Original Survey and Land Subdivision* (Chicago: Rand McNally, 1966), pp. 84 f.

[20] G. Schwarz, *Allgemeine Siedlungsgeographie* (Berlin: de Gruyter, 1961), pp. 158–205.

Niemeier finds in the vicinity of Buenos Aires a similarly varied rectangular network interrupted by long lots; and Künzler-Behncke, like others before him, notes in the Po Valley the tenacity of the rectangular system on a continent better known for its irregular farm and field patterns.[21]

The Politicized Landscape

Still another approach to the role of politics in the agricultural landscape is to consider a particular landscape in terms of all related political developments, rather than to study a particular political event in the light of its influence on the landscape. This reverse approach has taken two paths: the study of a boundary zone; and the consideration of a landscape in a country whose farming economy is strictly controlled by the government.

Contrasts between the farming complex on one side of a boundary and that on the other express themselves in a variety of ways, a situation that is reflected even in the meager reports so far made on the subject. James commented on the role of American tariffs in keeping agriculture in the Mexican part of the Imperial Valley less developed than in the American portion. A similar contrast in land use intensity along the boundary between Queensland and New South Wales has been described by Rose, but in this case it is explained more as a result of the greater vigor of the Queensland Government in settlement policy. Verhasselt found more Belgian proprietors across the boundary in the southwestern Netherlands than Dutch on the Belgian side, a finding which he associates with the cheaper land of the Dutch strip, its nearness to the rich industrialized cities of Belgium, and its earlier control by Belgium. Watson finds the contrasts between American and Canadian rural settlement patterns in the Red River Valley at least as striking as the similarities in types of land use. The significant role of the French in Canadian settlement is given as the main reason for these differences. And along what is probably the most studied border zone from the agricultural point of view, that shared by Germany and the Netherlands, geographers like Winterberg and Schwind see the long life of the boundary as the principal force behind the wealth of differences in an otherwise remarkably similar physical environment.[22]

[21] D. W. Meinig, *On the Margins of the Good Earth* (Chicago: Rand McNally, 1962), pp. 96–99; G. Niemeier, *Siedlungsgeographie* (Braunschweig: Westermann, 1967), p. 116; R. Künzler-Behncke, "Das Zeturiatsystem in der Po-Ebene," in *Festschrift Frankfurter geographische Gesellschaft 1836–1961* (Verein für Geographie und Statistik zu Frankfurt a. Main) (Frankfurt am Main: Kramer, 1961), pp. 159–70.

[22] P. E. James, *Latin America* (New York: Odyssey, 1942), pp. 623–34; A. J. Rose, "The Border Between Queensland and New South Wales," *Australian Geographer,* Vol. 6 (1955), 3–18; Y. Verhasselt, "Frontière politique et structure agraire: L'exemple de la Flandre zélandaise," *Études rurales,* Vol. 12 (1964), 95–108; Watson, *op. cit.;* A. Winterberg, *Das Bourtanger Moor. Die Entwicklung des gegenwärtigen Landschaftsbildes und die Ursachen seiner Verschiedenheit beiderseits der deutschholländischen Grenze* (Kiel: Geographisches Institut, Univ. Kiel, 1957); M. Schwind, *Deutsch-Niederländische Begegnung im Raum dès Bourtanger Moors* (Kiel: Hirt, 1962).

The most impressive evidence of political authority on the rural landscape—in the socialist countries of Eastern Europe, the Soviet Union, and China—has nowhere received the attention its magnitude deserves. Where the transformation of agriculture by government has been most far-reaching, in the Soviet Union, this lag may be laid to the smaller amount of attention paid to economic geography than to physical geography. The same favoritism can be seen in the heavy emphasis put on physical-geographic regionalization rather than on the actual agricultural picture. Since 1956, however, Soviet geographers have begun to shift their stress. There is a growing number of regional studies of the distribution of cropping systems, as by Kryuchkov on the Semipalatinsk Oblast, and a rapidly expanding list of atlases which provide a more comprehensive view of the agricultural complex, in particular the impressive *Agricultural Atlas of the U.S.S.R.*[23] Information about rural settlement patterns has been even slower in coming, primarily because settlement geography has up to now been of little interest to Soviet economic planners. This situation may now be remedied in light of the newer Party pronouncements on the necessity of equalizing the living conditions of village and city, the accomplishment of which will require extensive data on the individuality, structure, and distribution of rural settlements. That the possibilities of rich finds are great has been shown by those settlement studies that already exist, particularly those by Kovalev.[24]

Much the same deficiencies exist in the research in the other socialist countries, although there too the trend toward increased activity is very evident. This is true especially in Poland where, under the stimulus of Kostrowicki, extensive and detailed land use mapping has been carried on; the success of this work has also stimulated cooperative field work among several Eastern European countries, the first product of which has been the excellent volume, *Land Utilization in East-Central Europe: Case Studies.*[25] Intensive mapping is also the main theme of agricultural geography in China, although most of the results have yet to be published on an extensive scale and made available outside of the country. A rich future harvest in agricultural-geographic research in China seems as-

[23] V. G. Kryuchkov, "Problems of Agricultural Regionalization of Semipalatinsk Oblast," in *Materialy IV mezhvuzovskogo soveshchaniya po rayonirovaniyu dlya sel'skogo khozyaystva* [*Proceedings of the Fourth Inter-University Conference on Regionalization for Agriculture*] (Moscow: Moscow Univ. 1963), cited by A. N. Rakitnikov and V. G. Kryuchkov, "Agricultural Regionalization," *Soviet Geography: Review and Translation*, Vol. 7, No. 5 (1966), p. 55; M. I. Nikishov, ed., *Atlas sel'skogo khozyaystva SSSR* [*Agricultural Atlas of the U.S.S.R.*] (Moscow: Glavnoye Upravleniye Geodezii i Kartografii SSSR, 1960).

[24] Especially *Sel'skoe rasselenie* [*The Rural Settlement Net*] (Moscow, 1963), and "Typen ländlicher Siedlungen in der UdSSR," *Petermanns Geographische Mitteilungen*, Vol. 101 (1957), 152–57.

[25] The land mapping is concisely described in J. Kostrowicki and W. Tysgkiewicz, *Land Use Studies in East-Central Europe*, Dokumentacja Geograficzna, Zeszyt 3 (Warsaw: Polish Academy of Sciences, Institute of Geography, 1968), pp. 43–60; J. Kostrowicki, ed., *Land Utilization in East-Central Europe: Case Studies* (Warsaw: Polish Academy of Sciences, Institute of Geography, 1965).

ured, however, if one can judge by reports of the numerous goals now being pursued and the few accounts of foreign investigators on the variety and dynamic aspects of the Chinese agricultural scene.[26]

[26] S. D. Chang, "The Role of the Agricultural Geographer in Communist China," *Professional Geographer*, Vol. 18 (1966), 125–28; Y. P. Chung, "Geography and Agricultural Development in China," *ibid.*, Vol. 20 (1968), 163–66; R. Dumont, *Chine surpeuplée* (Paris: Esprit, 1965); G. Fochler-Hauke, "Die chinesischen Volkscommunen," *Geographische Rundschau*, Vol. 18 (1966), 137–44.

CHAPTER SEVEN

The historical context

The Search for Beginnings

As concrete and contemporary as the agricultural landscape is, it is impossible to view it without something of a historical perspective. Its development, its still shifting patterns, and its hints of future alterations have attracted a wide range of scholars. Nowhere is this allure more evident than in the baffling diversity of opinions offered on the possible sites of the first agriculture. Geographers, ethnologists, geneticists, botanists, historians, and anthropologists are only some of the many specialists who have become interested in the subject, the resulting disagreements, according to some, becoming more of a hindrance than a help in progress toward discovery. Nevertheless, some basic lines of argument can be discerned. The largest block of opinion, represented by such workers as the culture historian Childe and the anthropologist Braidwood, favors the river valleys and loess lands of the Iranian-Mesopotamian area as the primary agricultural hearth. Another view, typified especially by the geographers Sauer and Wissmann, is that such areas were too liable to flood and drought and that the more moist and botanically diverse hill and mountain lands of tropical Southeast Asia were the original center. Both groups also have different opinions of what type of crop cultivation was first to appear, the larger group favoring the sowing type; the smaller, the planting type.[1]

[1] V. G. Childe, *Man Makes Himself* (London: Watts, 1937); R. J. Braidwood, *The Near East and the Foundations for Civilization* (Eugene: Oregon State System for Higher Education, 1952); C. O. Sauer, *Agricultural Origins and Dispersals* (New York: American Geographical Society, 1952); H. V. Wissmann *et al.*, "On the Role of Nature and Man in Changing the Face of the Dry Belt of Asia," in *Man's Role in Changing the Face of the Earth,* ed. W. L. Thomas, Jr. (Chicago: Univ. of Chicago, 1956), pp. 278–303.

The smaller group has been particularly criticized for its lack of archeological evidence. But Harris has defended their position by noting the faster rate of decomposition of organic materials in the humid tropics and the much less intense archeological research in Southeast Asia as compared with the Middle East. Smolla has also remarked that nobody has ever maintained that one can decipher archeologically the early stage of an agriculture based on crops of the planting type.[2]

Another and more basic objection to the thesis of a Southeast Asian agricultural hearth has been made by Helbaek and others; they question the assumption that areas of greater plant diversity are also more likely to be the original centers of crop plants, pointing out that plant variety may result from many causes, such as human migrations and cross-fertilization of plants, in addition to the antiquity of local agriculture.[3]

Opinions also diverge on the question of where herding originated. One view, advanced by Hahn before the turn of the century and still supported in its essentials by Sauer and others, is that herding began among grain farmers in Southwest Asia, who raised sheep and goats. A variant is that herding originated among areas of grain culture, but that the principal animal was the cow. Schmidt and a few others, however, deny the association of herding with grain cultivation and believe that this economic form started with a reindeer-hunting population in the taiga. Narr believes that the archeological evidence is still insufficient to establish whether livestock raising originated in a pure form or with grain farming.[4]

Other basic disagreements exist about the kinds of routes followed by plants and animals as they were dispersed from their agricultural hearths. The advocates of the primacy of the Southeast-Asian hearth naturally assign a pattern of movements for that area that is largely unacceptable to supporters of the Southwest Asian hearth, although most of the patterns proposed for other areas by those who believe in the Southeast Asian hearth, e.g., Sauer, are accepted by many in the opposing camp. Sauer and several other geographers also reject the widely held belief that similar plants and cultivation practices originated independently in different areas, particularly in the New and Old Worlds; they suggest that there was a diffusion of planting cultures from a common tropical Asian base. Sauer feels that the New World was exposed to dispersal routes across both the northern and central Pacific, although he does not indicate them on his maps (combined in Fig. 5). More

[2] D. G. Harris, "New Light on Plant Domestication and the Origins of Agriculture: A Review," *Geographical Review,* Vol. 57 (1967), 96; G. Smolla, "Zum Problem der Enstehung der Landwirtschaft," *Zeitschrift für Agrargeschichte und Agrarsoziologie,* Vol. 12 (1964), 8.

[3] Reviewed in R. O. Whyte, "Evolution of Land Use in South-western Asia," in *A History of Land Use in Arid Regions,* ed. L. D. Stamp (Paris: UNESCO, 1961), pp. 86 f.

[4] E. Hahn, *Die Haustiere und ihre Beziehungen zur Wirtschaft des Menschen* (Leipzig: Duncker & Humblot 1896); Sauer, *op. cit.,* pp. 84 f.; P. W. Schmidt, "Zu den Anfängen der Herdentierzucht," *Zeitschrift für Ethnologie,* Vol. 76 (1951), 1–40, 201 ff.; K. J. Narr, "Early Food-Producing Populations," in *Man's Role in Changing the Face of the Earth,* pp. 136 f.

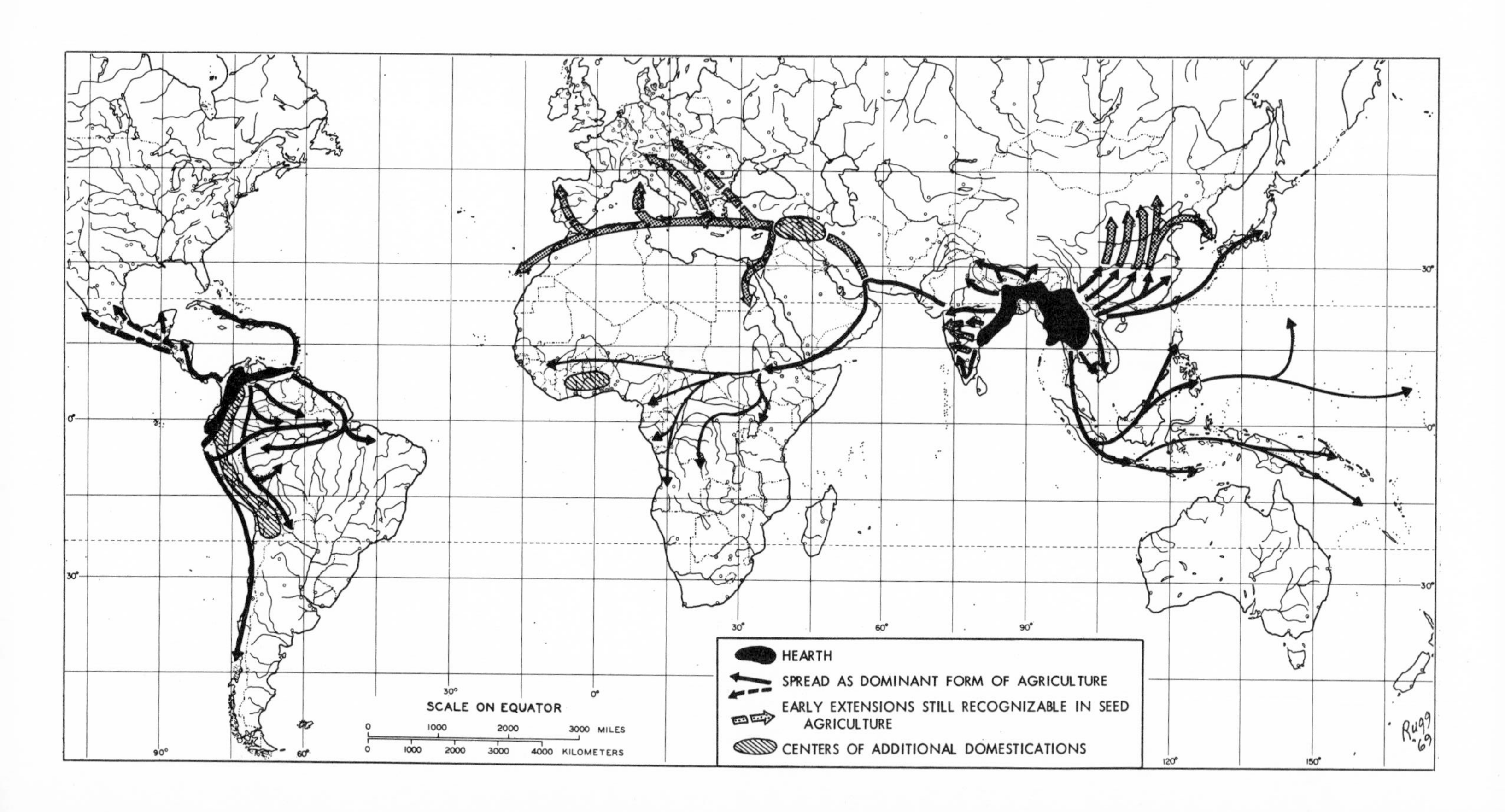
HEARTH
SPREAD AS DOMINANT FORM OF AGRICULTURE
EARLY EXTENSIONS STILL RECOGNIZABLE IN SEED AGRICULTURE
CENTERS OF ADDITIONAL DOMESTICATIONS
SCALE ON EQUATOR
0 1000 2000 3000 MILES
0 1000 2000 3000 4000 KILOMETERS
Rugg '69

Figure 5. *Origin and dispersal of planting, according to C. O. Sauer. [Adapted from plates I and II of* Agricultural Origins and Dispersals *(1952), by permission of the author and the American Geographical Society.]*

recently, MacNeish has greatly reinforced the argument for parallel invention with his discoveries in Mexico of plant remains dated, by radiocarbon, as far back as 7000 B.C., approximately the same time as that determined for the oldest remnants of plants found in the Old World.[5]

Naturally, much less controversy has accompanied attempts to trace the general movements of plants and animals and associated farming methods since the beginning of modern times. The start of modern agricultural movements on a large scale is associated with the flood tide of Europeans that began to spread over the earth in the sixteenth century. Otremba, among others, has described the Europeanization of the landscape; he also notes two subsequent movements, a return flow to Europe of plants native to the New World and a continuing movement of plants within the tropics, some of them penetrating the subtropics and a few even the higher middle latitudes. Otremba also admits of additional flows, such as the spread of the East Asian mulberry, soybean, and rice, but he does not consider them equal in magnitude and variety to the other three movements.[6]

The other side of the story of plant and animal dispersal, how these movements affected a receiving landscape in its totality, has received much less thought. The account by Clark of the massive transformation of the vegetative cover in South Island, New Zealand, by foreign imports is a classic among the very few works of this kind.[7]

The search for the origin and dispersal of the more structural characteristics of the agricultural landscape, its field and settlement forms, has been just as vigorous as the search for the production units, plants and animals. But this search has also been severely hindered by the more human character of these institutions and by all the rapid and frequently irrational changes in morphology that this characteristic implies. Intimate knowledge of local history is thus vital; consequently, most of the work on this subject, although intense, has been drastically restricted. And even for these restricted areas, the original centers and dispersal routes of particular field and settlement forms have usually not been fully clarified.

In some cases, regional findings have at first shown great promise of providing a principle by which origins and movements could be

[5] Sauer, *op. cit.*, pp. 54–61; R. S. MacNeish, "Ancient Mesoamerican Civilization," *Science*, Vol. 143 (1964), 531–37.

[6] E. Otremba, *Allgemeine Agrar- und Industriegeographie*, 2d ed. (Stuttgart: Franckh, 1960), pp. 141–48.

[7] A. H. Clark, *The Invasion of New Zealand by People, Plants and Animals: The South Island* (New Brunswick, N.J.: Rutgers Univ. Press, 1949).

deduced for regions well beyond the study area; yet all such discoveries have eventually proved useless in this respect. The possibility that particular settlement forms may have begun and spread through a particular physical environment, for example, has long been implied. Some geographers have seen relief as the key, finding farm villages primarily on the plains and isolated farmsteads on rougher terrain because of the greater fragmentation of arable land there; some have felt that soil condition is most determining, viewing farm villages as concentrated in areas where swampiness and the danger of flood encourage grouping for mass protection; still others have felt that the distribution of water is the primary determinant of settlement form, finding farm villages more common in the dry areas where water sources are few and widely dispersed and the majority of isolated farmsteads where water is abundant and uniformly distributed. None of these correlations is without serious exceptions, however, as Demangeon, himself a geographer, has so well shown.[8]

Correlations of settlement forms with agents other than the physical environment have likewise produced no satisfactory bases for eventually constructing a world pattern of origins and dispersals. The probability that these distributions could be based on movements of ethnic groups has been dashed by the subsequent criticism of Meitzen's prodigiously detailed work. He attributed the striking contrast between the farm villages and isolated farmsteads of Western and Central Europe to an original difference in ethnic makeup, the villages belonging to Germanic peoples and the isolated farmsteads to Celts. Among the more recent exceptions to this ethnic theory are Ilešič's findings that differences between Slavic and Germanic areas of early colonization in Slovenia are indistinguishable in the landscape, and Thrower's conclusion, in a study of two townships in Ohio, that no definite association can be established between the amount of regularity in settlement patterns and differences in cultural traits.[9]

Many of the exceptions made to Meitzen's theory have been based on the supposed universality of economic drives, yet here, too, reservations exist. For even if the pattern of the spread of farming systems is known, there is no guarantee that changes in agricultural practices evoke the same changes in settlement form in all areas. Demangeon has shown that periodic redistribution of land as a means of providing new land for younger generations has encouraged agglomerated settlement in some countries and left dispersed settlement in other countries intact.[10]

The quest for the oldest field form, block or strip, and its origin

[8] A. Demangeon "La géographie de l'habitat rural," *Annales de géographie*, Vol. 36 (1927), 9–13.

[9] A. Meitzen, *Siedlung und Agrarwesen der Westgermanen und Ostgermanen, der Kelten, Romer, Finner und Slaven*, 4 vols. (Berlin: Besser, 1895), Vol. 1, p. 520; M. S. Ilešič, "Les problèmes du paysage rural en Yougoslavie nord-occidentale, et spécialement en Slovenie," *Géographie et histoires agraires*, Annales de l'est, Mémoire no. 21 (Nancy: Univ. de Nancy, 1959), pp. 290 f.; N. J. W. Thrower, *Original Survey and Land Subdivision* (Chicago: Rand McNally, 1966), pp. 29 f.

[10] Demangeon, *op. cit.*, pp. 19–21.

has been somewhat more rewarding than the search for similar information on farm habitations, where variations are far more numerous and noneconomic motivations much more influential. Intense archeological work, especially by Hatt, points to the primacy of the irregular block field form in Europe and also to the possibility that it was introduced from the south, thus hinting of an origin synonymous with the agricultural hearth. This evidence, plus the generally accepted thesis that hoe culture is older than plow culture and the readily observable fact that irregular blocks dominate the hoe culture—largely tropical—landscape today, have convinced Niemeir that this field form is the oldest. Whether this form originated in Southeast or Southwest Asia is still problematical, for hoe culture was carried on in both places in prehistoric times.[11]

The inability of researchers to uncover material that could be a basis for a unitary theory of centers of origin and routes of dispersal of field and settlement forms should not blind one to the information that has been obtained and its value in the investigation of still other problems. In ascribing a diffusing role to the long-lot survey system, for example, Deffontaines has shown that the cultural background of French settlers, as well as their practicality, explains the close relationship of the long lot to North American waterways. Cultural motives are assigned even greater importance than practical considerations by Zelinsky, who found no evidence of diffusion of the New England connecting barn into adjoining areas with equally severe winters. In contrast, Kniffen found the climate a central force in the changing size and function of the Pennsylvania barn as it moved southward.[12]

The Manner of Change

Another major sign of interest in the historical aspects of agricultural geography is the variety of historical approaches used. Some writings have emphasized stages, one group leaning more toward their historical character, namely, the type of sequence, and another group emphasizing more their effects on a particular area. Still others have considered phenomena in terms of a long, relatively undifferentiated development, a particular historical period, or current changes.

Certainly the most polemical research has been that on the specific chronology of agricultural forms. Hahn deserves primary credit for the first major reorientation of thinking on the long-accepted sequence of hunting-herding-farming. Pointing to cases where hunting-collecting and plant cultivation exist side by side, Hahn noted that the form of cultiva-

[11] G. Hatt, *Oldtidagre* (København: Munksgaard, 1949); G. Niemeier, *Siedlungsgeographie* (Braunschweig: Westermann, 1967), p. 53.

[12] P. Deffontaines, "Le rang, type de peuplement rural du Canada français," *Cahiers de geographie*, Vol. 5 (1953), 27–29; W. Zelinsky, "The New England Connecting Barn," *Geographical Review*, Vol. 48 (1958), 553; F. Kniffen, "Folk Housing: Key to Diffusion," *Annals of the Association of American Geographers*, Vol. 55 (1965), 563 f., 574.

tion was always hoe culture. He thus revised the sequence to one of hunting-hoe cultivation-plow cultivation; herding, he also claimed, derived from, rather than preceded, cultivation. Although these views are now accepted by the majority, objections have become increasingly pointed, particularly those by Leser and Werth. Both claim that plow cultivation developed from the "foot plow" instead of the hoe, Werth further contending that plow cultivation has not replaced hoe cultivation but has been added to it. Schmidt also denies that herding developed as an offshoot of cropping, claiming gathering or collecting as its source form; Hatt believes that herding was originated by both hunters and agriculturists.[13]

The other approach to stages in agriculture has concerned itself with the changes a specific region has undergone. The assumption that definite stages can be recognized, during each of which the agricultural use of an area remains more or less constant in its fundamental aspects, is an old and still popular one among European geographers. For a similar situation among American geographers, one must credit in particular the stimulus of Whittlesey's article on "sequent occupance." Users of this approach, however, still have varying perspectives. Some have considered the changes affecting only one crop or form of economy, such as Perpillou in his close study of "a century of agricultural evolution" in the French vineyards of Languedoc and Rousillon, and McCune in his article on the "sequence of plantation agriculture" in Ceylon. Others—and they have been more numerous—have looked more upon the changing regional complex of agricultural patterns. One of the most recent, and certainly most detailed, illustrations of this view is the presentation by Tisowsky of the four historical periods marking agricultural change in the upper portion of the German county of Rheingau since the middle of the eighteenth century. In the New World, Hoy has used this approach toward "land use periods" in Guadeloupe, the French West Indies. How fruitful such work can be when carried on in much larger areas has been established by Meinig, who used stages to explain the changing ranks of agricultural regions in the Western United States.[14]

[13] E. Hahn, "Waren die Menschen der Urzeit zwischen der Jägerstufe und der Stufe des Ackerbaus Nomaden?" *Das Ausland,* Vol. 64 (1891), 481–87, cited by F. L. Kramer, "Eduard Hahn and the End of the 'Three Stages of Man,'" *Geographical Review,* Vol. 57 (1967), 84. Kramer summarizes the objections of Leser, Werth, and others to parts of Hahn's theories, *op. cit.,* pp. 86–89. See also P. W. Schmidt, *op. cit.,* and E. C. Curwen and G. Hatt, *Plough and Pasture* (New York: Schuman, 1953; Collier Books Edition, 1961), p. 232.

[14] D. Whittlesey, "Sequent Occupance," *Annals of the Association of American Geographers,* Vol. 19 (1929), 162–65; A. V. Perpillou, "Un siècle d'évolution agricole dans les vignobles français du Languedoc et du Roussillon," *Wiener geographische Schriften* (Festschrift Leopold G. Scheidl zum 60. Geburtstag, eds. H. Baumgartner *et al.,* Part I), Nos. 18–23 (1965), pp. 268–78; S. McCune, "Sequence of Plantation Agriculture in Ceylon," *Economic Geography,* Vol. 25 (1949), 226–35; K. Tisowsky, "Ackerland, Rebflächen und Obstkulturen im oberen Rheingau," in *Festschrift frankfurter geographische Gessellschaft 1836–1961,* Verein für Geographie und Statistik zu Frankfurt a. M. (Frankfurt am Main: Kremer, 1961), pp. 389–423; D. R. Hoy, "Changing Agricultural Land Use on Guadeloupe, French West Indies," *Annals*

UNIVERSITY OF PITTSBURGH LIBRARY
AT BRADFORD

Despite the popularity of the concept of stages, even more writers have seen the historical role as not involving a step-like progression. For some, change has been a long, comparatively constant process, as for Dion, who traces the centuries-old spread of the vine through France, and for Raup, who views the traditionally cited "livestock-grain-fruits and vegetables" sequence for California agriculture rather as a general intensification process. For others, the most interesting historical aspect of the agricultural geography of an area has been the situation during a single historical period. Older and classic works, such as Semple's on ancient agriculture in the Mediterranean area, are echoed in modern studies like Thomas's on Welsh agriculture during the Napoleonic Wars. Not surprisingly, many of the geographers interested in this "snapshot" view of past agricultural geographies have also been attracted by the role of agriculture in colonization. Among North American geographers, the first major effort of this kind was Brown's attempt to picture the agricultural geography of each major region in the United States as it was opened to settlement. Studies like Gentilcore's, on general agricultural conditions in the Vincennes, Indiana, area during the early stages of colonization, and Meinig's, on farm beginnings in South Australia, continue this focus.[15]

Still another research trend in the study of the historical aspects of agricultural geography is represented by the massive volume of writings on contemporary changes. Advances and retreats of individual crops continue to be the most commonly observed of these changes. The rapid expansion of soybeans in the American Corn Belt has already prompted Haystead and Fite to rename the region the Corn-Soy Belt. Similarly, the accelerating retreat of cotton in the Cotton Belt has inspired Prunty and others to reject the "Belt" part of that regional title. The author has noted the increase in southern California of staple products raised for the burgeoning local market at the expense of the more specialized products destined for distant markets, a development now seriously qualifying the traditional labeling of the area as primarily an agricultural specialty center. Nor are similar examples of sizable crop shifts lacking in the well-established patterns of Northwestern Europe, as Hartke shows in his

---

of the Association of American Geographers, Vol. 52 (1962), 441–54; D. W. Meinig, "The Growth of Agricultural Regions in the Far West: 1850–1910," *Journal of Geography,* Vol. 54 (1955), 221–32.

[15] R. Dion, *Histoire de la vigne et du vin en France des origines au XIXe siècle* (Paris: Privately published, 1959); H. F. Raup, "Transformation of Southern California to a Cultivated Land," *Annals of the Association of American Geographers, Supplement,* Vol. 49 (1959), 58–78; E. S. Semple, "Ancient Mediterranean Agriculture," *Agricultural History,* Vol. 2 (1928), 61–98, 129–56; D. Thomas, *Agriculture in Wales during the Napoleonic Wars* (Cardiff: Univ. of Wales, 1963); R. H. Brown, *Historical Geography of the United States* (New York: Harcourt, Brace, 1948); R. L. Gentilcore, "Vincennes and French Settlement in the Old Northwest," *Annals of the Association of American Geographers,* Vol. 47 (1957), 285–97; D. W. Meinig, *On the Margins of the Good Earth* (Chicago: Rand McNally, 1962).

account of the push of hybrid corn into northern France and the southward retreat of vineyards from the same region.[16]

Where crop changes have been of sufficient magnitude, regional agricultural hierarchies have also been seriously altered or changed. These developments, too, have been noticed. Enyedi has disclosed that an increasing shift from cereals to industrial crops, vegetables, fruits, and the vine since the mid-1930's has enabled the Balkan countries to reduce the gap in agricultural productivity between them and other Eastern European nations. An example of a widening regional gap is offered by Hart in his study of changes in sheep distribution in Britain from 1938 to 1951. His many detailed maps document a reduction in sheep over most of the nation, but particularly in the lowlands, where much land is being transferred to grain and fodder crops. During approximately the same period, from 1939 to 1954, this writer observed a change in regional leadership within California in agricultural intensity as measured by the value of production per crop-acre: the continuing urbanization of the coast and a shift toward intensive livestock raising in the interior dropped southern California's agricultural intensity to less than that in the oases of the southeastern deserts.[17]

Another type of current agricultural change that is attracting notable research interest is the increasing thrust of certain forms of farming into environments previously considered hostile. The proliferation of dairy farms in various parts of the tropics and subtropics has already been widely reported.[18] In turn, the plantation, an institution considered by most as generally synonymous with the tropics, has moved into cooler latitudes. Gerling's argument that many large farming operations in the subtropics are just as large and specialized as tropical plantations has received support from Prunty's studies of large farms in subtropical southeastern United States, and Gregor claims, on the basis of the swift rationalization of farming operations in the United States, that plantations may now be considered to be well distributed in several parts of the higher middle latitudes as well.[19] Technological improvements in farming, particularly the expanded use of large machines, have impressed Faucher, in this case by making possible widespread advances of dry

[16] L. Haystead and G. C. Fite, *The Agricultural Regions of the United States* (Norman: Univ. of Oklahoma, 1955), pp. 140–61; M. C. Prunty, Jr., "Recent Quantitative Changes in the Cotton Regions of the Southeastern States," *Economic Geography,* Vol. 27 (1951), 189–208; H. F. Gregor, "Urbanization of Southern California Agriculture," *Tijdschrift voor Economische en Sociale Geografie,* Vol. 54 (1963), 276; W. Hartke, *Frankreich,* 2d ed. (Frankfurt/Main: Diesterweg, 1966), pp. 65–69.

[17] G. Enyedi, "The Changing Face of Agriculture in Eastern Europe," *Geographical Review,* Vol. 57 (1967), 366–68; J. F. Hart, "The Changing Distribution of Sheep in Britain," *Economic Geography,* Vol. 32 (1956), 260–74; H. F. Gregor, "Regional Hierarchies in California Agricultural Production, 1939–1954," *Annals of the Association of American Geographers,* Vol. 53 (1963), 30–33.

[18] Much of their work is summarized and analyzed in H. F. Gregor, "Industrialized Drylot Dairying: An Overview," *Economic Geography,* Vol. 39 (1963), 299–318.

[19] These and other views of the plantation are reviewed in *idem,* "The Changing Plantation," *Annals of the Association of American Geographers,* Vol. 55 (1965), 221–38.

farming at the expense of ranching.[20] The attention paid by writers to the advances of irrigation into the dry lands is only mentioned here, some of the more important articles having been noted in the previous chapter.

Lag and Retrogression

Although the larger amount of work done in agricultural geography within a historical context has been on change that is emphatic and reflects intensification, significant interest has also been shown in those aspects of agriculture which exhibit inertia or even retrogression. One focus has been on the traditional lag of farming systems and farm morphology behind economic change; another has been on the abandonment of farm settlements and farmland. European geographers have of course been particularly interested in these negative historical aspects, for their rural landscapes bear the stamp of a long tradition. Opinions from across the Atlantic are not missing, however.

The often sharp contrast between the farming system and modern economic demands has especially interested the historically oriented French geographers. Some have been so impressed with the inertia of farming that they have suggested it has prehistoric roots, the very ancientness of customs helping to make them permanent routines of the peasant. Others deny such permanence, claiming that the bewildering medley of farming practices is less archaic than an evidence of constant change, albeit at varying speeds, and adjustment to new economic requirements.[21] Faucher's opinion, somewhere between, is that rural cultures pass alternately through periods of "calm" and "invention." Those of invention are times of intellectual effervescence, marked by new ideas and experiments which usually originate outside the peasantry; those of calm are times of "stiffening," when farming systems enter into folklore, an "ensemble of facts which tradition transmits with a cortege of explanatory reasons accepted without criticism." To the complaint that such a cyclical pattern can occur only among closed economies, Faucher has replied that no farming economy is completely closed and that even in very open economies the pattern can be traced.[22]

The disparity between the morphology of the farm and current economic requirements has engaged the thought of researchers even more, for farms in many parts of the world have become increasingly anachronistic through repeated parceling among inheritors at the same time that agricultural technology has been putting an ever greater premium on larger and more compact farms. Although much of the writing on this problem has dwelt on the progress made by farm-consolidation programs (noted in Chapter 6), enough has also been written to indicate that the forces working against speedy rationalization are

[20] D. Faucher, *Le paysan et la machine* (Paris: Editions de Minuit, 1954), pp. 127–40.

[21] E. Juillard, "La géographie agraire," *La géographie française au milieu du XXe siècle* (Paris: Baillière, 1957), p. 166.

[22] D. Faucher, "Réflexions sur la méthode en géographie agraire," *Les études rhodaniennes,* Vol. 21 (1946), 90–92.

still formidable. Even in the American Midwest, where advanced agricultural technology is more revered than in most other parts of the world, Zelinsky has observed that the consolidation and enlargement of farms have not kept pace with that in many other sections of the country. He suggests, as causes, the unusually high cost of land and buildings, the relative social stability of the family farms, and the economic alternatives for farm residents in nearby cities. The only remaining solution, therefore, has been to expand the farm through purchase of noncontiguous plots. Such "rational fragmentation" has its drawbacks, however, the distance between plots posing much the same problem that the farmer faces in older agricultural areas. Fragmented farms of this type have already been reported from such widely scattered areas as the Middle Atlantic states, Georgia, England, California, Wisconsin, the Great Plains, the Philippines, and British Columbia.[23] Meanwhile fragmentation of original properties also continues, even in many parts of the United States, particularly where land costs are low and the desire for a rural residence by city workers is great.

The lag in changes in farm morphology has also been of concern to those who are attempting to discover if any obligatory relationships exist between particular field forms and certain aspects of the physical or human environment. Such correlations have been made extremely difficult by the complexity of the field mosaic, a condition resulting from the historical overlapping brought about by the lag in field changes and the variation in its rate from place to place. Attempts to relate particular field forms to individual elements such as terrain, use, farm system, and farm tools have so far failed to produce conclusions free of significant exceptions, although the relationship between large and squarish fields and modern irrigation and mechanized farming is accepted by some as standard.[24] Because field outlines change so much, Otremba has suggested that they be viewed more as changing, and less as static, forms. This dynamic condition, he maintains, is basically due to changing social conditions, a view now held by a majority of geographers. Some, however, in particular Krenzlin, argue for the primacy of economic motives.[25]

[23] W. Zelinsky, "Changes in the Geographic Patterns of Rural Population in the United States: 1790–1960," *Geographical Review,* Vol. 52 (1962), 522. The expansion of the process of fragmentation of farms in British Columbia is shown cartographically in A. Krenzlin, *Die Agrarlandschaft an der Nordgrenze der Besiedlung im intermontanen British Columbia* (Frankfurt am Main: Kramer, 1965), pp. 29–31.

[24] What may be the exception that proves the rule is J. F. Hart's recent finding of a close functional relationship between both field size and form and farming type in central Indiana, an agricultural area comparatively young by world standards, reported in "Field Patterns in Indiana," *Geographical Review,* Vol. 58 (1968), 460–71. G. Schwarz gives a succinct account of the various attempts to correlate field forms with particular elements: *Siedlungsgeographie,* 3d ed. (Berlin: de Gruyter, 1966), pp. 248–54.

[25] E. Otremba, *Die Entwicklungsgeschichte der Flurformen im oberdeutschen Altsiedelland, "Berichte zur deutschen Landeskunde,"* Vol. 69 (Bad Godesburg: Bundesanstalt für Landeskunde und Raumforschung, 1951), pp. 363–81; A. Krenzlin, "Blockflur, Langstreifenflur und Gewannflur als Ausdruck agrarischer Wirtschaftsformen in Deutschland," *Géographie et histoire agraires,* "Annales de l'est," Mém. 21 (Nancy: Faculté des Lettres et des Sciences humaines, L'univ. de Nancy, 1959), pp. 353–69.

A further problem facing those interested in field forms as indicators of particular economic or social conditions is that farm-ownership patterns lag even behind changes in field patterns. Hartke has criticized Krenzlin's argument for economic motivations because much of the research was based on cadaster records, unfortunately usually the only maps of very old land use patterns available. Brunet has shown that even today the widespread leasing of lands in the Paris Basin has produced a far simpler crop pattern than the cadaster sheets suggest. Somewhat the same conclusion has been reached by Kaups and Mather, who discovered that farm properties in northern Michigan have remained small, although operational units have been enlarged by renting.[26]

Still another variation in the rate of change of farm morphology occurs with the persistence of the settlement form despite widespread changes on the rest of the farm. This situation has been widely reported of both farm villages and individual farmsteads. Some of the most striking examples of the tenacity of the village have been reported in Northwestern Europe, where cultural attachment to this kind of settlement form has long been close. Evers, for instance, has shown how villages in northwestern Germany have maintained themselves in the face of extensive consolidation of fields; while Bell, on the basis of his findings in northeastern France (Alsace), believes that attachment to village life is so strong that consolidation of fields will probably never be completely accomplished. But even in younger lands, where fields have often become large and geometrical and farming practices highly rationalized, farm villages have been able to survive, as Warkentin's account of the Mennonite settlements in the Manitoba wheatlands so beautifully illustrates. Economic advantage, on the other hand, may be the very reason for the survival of agglomerated settlements. Such is the case, as Chardon shows, on the henequen haciendas of northwestern Yucatan, whose possession of most of the processing plants and closer integration of cultivation and management have allowed their building complexes to thrive despite the subdivision of most hacienda land into ejidos.[27]

The similar persistence of smaller farmsteads and of their original characteristics in the face of field changes has also been frequently described, often in terms of disparity between form and function. As

[26] W. Hartke, "Eine Geographie des modernen, hochproduktiven Fruchtwechselbetriebes," *Erdkunde*, Vol. 16 (1962), 148; P. Brunet, *Structure Agraire et Économie Rurale des Plateaux Tertiaires entre la Seine et l'Oise* (Caen: Caron, 1960), pp. 61–83; M. Kaups and C. Mather, "Eben: Thirty Years Later in a Finnish Community in the Upper Peninsula of Michigan," *Economic Geography*, Vol. 44 (1968), 64, 68.

[27] M. W. Evers, "Agrarlandschaft und bäuerliche Siedlung; Zusammenhänge und Probleme, erläutert am Beispiel des Gestaltungsprinzips einer frühmittelalterlichneuzeitlichen Siedlungsform," *Géographie et histoire agraires*, "Annales de l'est," Mém. 21 (Nancy: Faculté des Lettres et des Sciences humaines, L'univ. de Nancy, 1959), p. 142, fn. 2; T. A. Bell, "The Alsatian Peasant and His Land," *Yearbook of the Association of Pacific Coast Geographers*, Vol. 28 (1966), 94; J. Warkentin, "Mennonite Agricultural Settlements in Southern Manitoba," *Geographical Review*, Vol. 49 (1959), 342–68; R. E. Chardon, "Hacienda and Ejido in Yucatan: The Example of Santa Ana Cucá," *Annals of the Association of American Geographers*, Vol. 53 (1963), 179 f., 186.

economic requirements change and specialized buildings lose their value, farmers commonly attempt to salvage their investment by continuing to use the structures, even though they may not be as well designed for their new function. Because changes in farming practices have been particularly rapid in the United States, it is only natural that references by American geographers to changes in building use are notably evident, albeit brief. In 1946, Durand was already observing that cheese factories in northwestern Illinois were being converted into barns for grain and hay, and in 1951 Gregor described the shift in barn use in southern California, from the housing of stock, grain, and feed to the sheltering of machinery and cars. A more recent example is the reference by Hart and Mather to the use of tobacco barns in the Onondaga district of New York as corn cribs.[28]

The functional conversion of farm buildings often appears to be a prelude to their abandonment, and may even itself be part of the decreasing intensity or abandonment of the farm unit. Quite often, too, the conversion of buildings to new uses on one farm is accompanied by many more abandonments of buildings on adjacent farms built originally for similar uses. The decreasing intensity or abandonment of farming operations has been viewed on all three morphological scales, the individual farm building, the farmstead, and the entire farm unit. References to abandonment of a particular type of building are extensive, applying to such diverse structures as chicken and brooder houses on a Montana wheat farm, tobacco barns in Kentucky and Tennessee, grape packinghouses in New York, "tank towers" on California farms, cheese factories in Wisconsin, and tenant houses on Southern plantations. Most such references, however, are only briefly made in the context of other research objectives.

The abandonment of the entire farmstead or farm has been scrutinized more purposively and for at least as long a period, the classic studies of Goldthwait, on the abandonment of farmsteads in New Hampshire, and of Baker, on the declining intensity of land use in the United States, going back to 1927 and 1929, respectively. Today, with the rural exodus stronger than ever, contributions of this type are especially numerous. Hart's account of the readvance of the forest in the Northeastern United States in the wake of land abandonment is one of the more penetrating works of this type. In Europe it is the work of Hartke on the increase of "social fallow" (*Sozialbrache*) in West Germany that especially stands out at present, although research on farm and land abandonment in medieval and early modern times has been traditional.[29]

[28] L. Durand, Jr., "Cheese Region of Northwestern Illinois," *Economic Geography,* Vol. 22 (1946), 36; H. F. Gregor, "A Sample Study of the California Ranch," *Annals of the Association of American Geographers,* Vol. 41 (1951), 278, fn. 9; J. F. Hart and E. C. Mather, "The Character of Tobacco Barns and Their Role in the Tobacco Economy of the United States," *ibid.,* Vol. 51 (1961), 278, fn. 9.

[29] J. W. Goldthwait, "A Town That Has Gone Downhill," *Geographical Review,* Vol. 17 (1927), 527–52; O. E. Baker, "The Increasing Importance of the Physical Conditions in Determining the Utilization of Land for Agricultural and Forest Production in the United States," *Annals of the Association of American*

Despite the different rates of change of the various parts of the agricultural landscape and the frequent fluctuations of those rates in any one area, geographers have continued to seek historical generalizations. These attempts have dealt with both the manner of change and the relation of one pattern of change to another. The well-known sign of a maturing agricultural landscape, the shift to ever more intensive land uses, was already being described inductively in 1926 by Baker and in 1929 by Rühl. To Baker, the most impressive aspect of the change was the increasingly closer correlation of the land use mosaic with the environmental detail; for Rühl, it was the many contrasts in "displacement energy" (*Verdrängungsenergie*) that supported a constant battle between the various land uses for areal superiority. Later, in 1953, Otremba became confident enough to express another generalization on the maturing rural landscape, that fields became smaller and less geometric. Another "rule" proposed by him was that all collective settlement begins with fields of marked regularity, while individual settlement leads in general to a net of irregular block forms.[30]

As to farmstead changes, most writers seem to agree that as the farm becomes more rationalized, houses become more standardized and less indicative of the function of the particular farm. This change has had most meaning for those studying the agricultural areas of the Old World, where local rural architecture has had a longer time to develop and become traditional, and where the combined house and barn is still common. Whether the other farmstead buildings are also becoming more standardized, as well as fewer in number, as many like Sorre have maintained, is less definite. Schwarz has cautioned that increasing mechanization and specialization can often engender the opposite developments, as they have on North American dairy farms.[31]

Much less has been theorized about the historical patterns of the most dynamic and elusive element of the agricultural landscape, its population. A notable exception is Zelinsky's claim to have recognized, at least in the United States, five distinct phases in the "life cycle" of rural population. The first three phases, rapid growth through in-migration, steady substantial growth largely through natural increase, and loss of population through out-migration, have already been experienced by most American counties; a fourth phase, stabilization or resettlement of the rural areas by a more or less urbanized population, has been taking place mostly in the Northeastern United States; and a fifth phase, a

---

Geographers, Vol. 11 (1929), 17–46; J. F. Hart, "The Three R's of Rural Northeastern United States, *Canadian Geographer,* Vol. 7 (1963), 13–22; K. Scharlau offers an excellent summary of the various research trends in Central Europe in "Sozialbrache und Wüstungserscheinungen," *Erdkunde,* Vol. 12 (1958), 289–94.

[30] O. E. Baker, "Agricultural Regions of North America," *Economic Geography,* Vol. 2 (1926), 479–88; A. Rühl, *Das Standortsproblem in der Landwirtschafts-Geographie* (*Das Neuland Ostaustralien*), Veröffentlichungen des Instituts für Meereskunde. Neue Folge. B—Historisch-volkswirtschaftliche Reihe, Heft 6 (Berlin: Mittler, 1929), pp. 121–24; Otremba, *Allgemeine Agrar- und Industriegeographie,* p. 157.

[31] M. Sorre, *L'habitat,* Vol. 3 of *Les fondements de la géographie humaine* (Paris: Colin, 1954), pp. 132 f.; Schwarz, *op. cit.,* pp. 107 f.

decrease in the recently revived rural population through loss of rural land to expanding cities, can be observed in a few metropolitan counties.[32]

The more complicated question of whether any orderliness exists in the way one aspect of the agricultural area changes in relation to another has also been answered in several ways. Differences between the growth rates of farm villages and isolated farmsteads have produced several recognizable sequences, according to French geographers like Derruau. Villages or isolated farmsteads may appear first (*concentration primaire, dispersion primaire*), only to be later challenged by an opposing settlement process (*dispersion secondaire, concentration secondaire*). This rival process may eventually even make its settlement form supreme as the original settlement type stagnates and disappears, the "final" settlement pattern thus being achieved through secondary dispersion or concentration *par substitution*.[33]

A variety of sequences has also been uncovered in the analysis of changing relationships between farmstead and field in the process of land abandonment. Mather found, in his study area near Kalamazoo, Michigan, that this process reflected itself in the landscape in several ways: (1) use of all the fields but with partial or no use of the farmstead; (2) use of all the farmstead but with some idle fields; and (3) use of none of the farm except for the occupied dwelling. About another facet of the settlement-field relationship, Glauert evidences the general opinion when he states that the tendency toward the isolated settlement increases as the farm, and thus also the fields, become larger. Chisholm, however, claims a long-term trend toward agglomerated settlement in spite of increasing farm consolidation. Using examples in England, the United States, and other advanced agricultural countries, he argues that shorter working hours and greater availability of transport are making it possible for farmers to travel farther and easier, and that at the same time the increased leisure is also creating greater social needs. Further, he adds, the conjunction of new production techniques in animal husbandry with transport improvements is making it increasingly possible to separate the land from the buildings.[34]

The Past as Prologue

The step from trend to outcome is short, so that although most workers in agricultural geography reject prediction as intellectually precarious at best, few also deny the worth of evaluating the chances for particular developments. Some, as we have seen, have been content to delineate trends and postulate cycles in such evaluations, but others have

[32] Zelinsky, "Changes in the Geographic Patterns," pp. 504 f.

[33] M. Derruau, *Précis de géographie humaine* (Paris: Colin, 1961), pp. 332 f.

[34] E. C. Mather, "One Hundred Houses West," *Canadian Geographer*, Vol. 7 (1963), 4; G. Glauert, "Agrargeographie," in *Allgemeine Agrargeographie*, ed. G. Fochler-Hauke (Frankfurt a. Main: Fischer Bücherei, 1959), p. 49; M. Chisholm, *Rural Settlement and Land Use* (London: Hutchinson, 1962), pp. 175–83.

been impelled by the peculiarly dynamic nature of the object to make forthright assessments of future possibilities. Changes in an agricultural landscape may be so numerous, rapid, and revolutionary, for example, that any treatment of such an area must necessarily be centralized on present and possible future changes. Certainly this has been the situation that has encouraged Prunty to forecast certain land use trends in the Southeastern United States, although for the nearest decade and "not without trepidation."[35] Still other researchers have been stimulated by the exotic character of their study object to speculate on the future. The unusual productivity of "drylot dairying" in the United States, despite its great contrast with most dairy farming in almost every major economic and physical characteristic, has been one of the latest stimulators.[36]

An increasing awareness of the usefulness of mathematical techniques has also begun to play a part in encouraging speculation about future agricultural changes. By assuming the same rate of agricultural development in the future as in the past, Enyedi has found it possible to calculate the number of years it would take for the differences among Eastern European countries in gross agricultural value per agricultural worker and per hectare of agricultural land to be eliminated. That the time needed for equalizing the productivity of the land proved to be much shorter than that needed for equalizing the productivity of labor has pointed up the much greater need in the future for massive capital investment in those countries that have up to now invested mostly in manpower. Bowden has found profitable the suggestion by Hägerstrand that theoretical models which simulate the process of adoption of innovations can "eventually make certain predictions achievable." Using the Colorado northern High Plains as his study area, Bowden first measured a known pattern of communication over space that could be used to simulate the adoption of irrigation, a pattern based on long-distance calls and attendance at a barbecue. From this pattern he constructed a probability model which he tested by simulating the past diffusion of wells and comparing the results with actual well patterns. Results proved favorable, so he then applied the model to a projection of diffusion of wells by 1975 and 1990. Another use of a model for predicting agricultural changes, this time in land utilization patterns from year to year, has been made by Henderson. The changes studied are those resulting from changing governmental policies and the reactions of farmers to differentials in returns.[37]

[35] Prunty, "Land Occupance in the Southeast: Landmarks and Forecast," *Geographical Review*, Vol. 42 (1952), pp. 460 f.

[36] Gregor, "Industrialized Drylot Dairying," pp. 314–18.

[37] Enyedi, *op. cit.*, pp. 369–72; L. W. Bowden, *Diffusion of the Decision to Irrigate* (Chicago: Univ. of Chicago, 1965), pp. 89–120; J. H. Henderson, "The Utilization of Agricultural Land: A Regional Approach," *Papers and Proceedings of the Regional Science Association*, Vol. 3 (1957), 99–114.

PART 3 the search for regions

CHAPTER EIGHT

Regional types and patterns

Single-Feature Regions

The regional context in agricultural-geographic research is even more evident than the historical, for the concern with areal variation has of necessity made such a framework mandatory. The majority of regional studies, as can readily be observed from the references cited so far, are restricted to a single region. Primary interest has been in objects within only a particular area, either because of the restricted distribution of phenomena, or the limitations of time, or both. But some writers have responded to the challenge of finding regional patterns in these phenomena or even of comprehending the total agricultural makeup of a region.

The regional patterns of single features have been contemplated the longest, both because of their basic character and their visibility. By 1883, Engelbrecht had sifted the data from the United States Census Bureau to construct individual distribution maps of major crops and animals. Later he did the same for other countries, and by 1930 he was able to summarize his findings in his world map of "agricultural zones" (*Landbauzonen*).[1] These zones were the areas occupied by sugar cane, cotton, or one of the principal grain crops, all of which were grouped into three larger "tropical," "subtropical," and "extratropical" zones. Today, the regionalization of crops and animals in the agricultural landscape is

[1] T. H. Engelbrecht, "Der Standort der Landwirtschaftszweige in Nord-Amerika," *Landwirtschaftliche Jahrbücher*, Vol. 12 (1883), 459–509; *idem, Die Landbauzonen der Erde*, Petermanns Mitteilungen, Ergänzungsheft 209 (Gotha: Haack, 1939).

carried out on a scale bigger than ever, and more of the effort is coming from specialists in governmental agencies.[2]

Academicians, and particularly geographers, however, have devoted increasingly less effort to single feature regions as their interest in commodities has lessened and their interest in more complex regional objects like farming types have grown. An exception to this trend has been the seeking of information about basic crop and animal distributions in areas where little is known about such patterns. Thus, some of the most recent and detailed single-feature regional stüdies made by individual geographers have been of cattle types, staple subsistence crops, and export commodities in Africa.[3] Whether efforts like these—particularly as interest in the resources and problems of underdeveloped areas continues to grow—will eventually slow the widening divergence between the two research groups is still uncertain. Otremba, in an agricultural geography text that stresses principles, has declared that single-feature regions are necessary to geographers only as a preliminary to synthesis. Pfeifer vehemently rejects this opinion, however, claiming that analysis of the distribution of a single feature is a thoroughly respectable and necessary scholarly activity in itself, and that the generalizations coming from such a work can seldom if ever be obtained through a study of complex combinations of factors.[4]

There is less cleavage of research orientation toward studies of elements of the agricultural landscape other than commodities. A major reason undoubtedly is the same one that has motivated geographers to continue their studies of crops and animals: the desire to know the agricultural-geographic patterns of a lesser-known area. Although considerable information about such things as work tools, field forms, and buildings is available, the multiplicity of subtypes and the complexity of their distribution have still prevented scholars from deriving the many and extensive regional patterns that are now so abundantly available for the agricultural commodities. Most of the writing has been by European geographers, although articles such as Marschner's, on field-form regions in the United States, and Hart and Mather's, on the regionalization of American tobacco barns, evidence an awakening interest elsewhere.[5] Studies of agricultural features other than commodities are also being

[2] The abundant and detailed maps which appear regularly in the Census of Agriculture's "Special Report," *Graphic Summary of Land Utilization* (Washington, D.C.: U.S. Govt. Printing Office) comprise one of many examples.

[3] W. Deshler, "Cattle in Africa: Distribution, Types, and Problems," *Geographical Review,* Vol. 53 (1963), 52–58; G. P. Murdock, "Staple Subsistence Crops of Africa," *ibid.,* Vol. 50 (1960), 523–40; W. A. Hance, V. Kotschar, and R. J. Peterec, "Source Areas of Export Production in Tropical Africa," *ibid.,* Vol. 51 (1961), 487–99.

[4] E. Otremba, *Allgemeine Agrar- und Industriegeographie,* 2d ed. (Stuttgart: Franckh, 1960), p. 47; G. Pfeifer, "Die allgemeine Agrar- und Industriegeographie von Erich Otremba in der 2. Auflage," *Erde,* Vol. 92 (1961), 154.

[5] F. J. Marschner, *Land Use and Its Patterns in the United States,* Dept. of Agriculture Handbook No. 153 (Washington, D.C.: U.S. Govt. Printing Office, 1959), pp. 8–31; J. F. Hart and E. C. Mather, "The Character of Tobacco Barns and Their Role in the Tobacco Economy of the United States," *Annals of the Association of American Geographers,* Vol. 51 (1961), 274–93.

made in ever larger numbers by census agencies and other more specialized governmental groups. In the United States, the contributions of that part of the Census Bureau producing the quinquennial Census of Agriculture have been particularly extensive.

Multiple-Feature Regions

From considering the regional pattern of one feature of the agricultural landscape to comparing that pattern with those of other features is but a short and logical step in the regionalization process. And as additional elements of the landscape continue to be uncovered, the taking of this step is ever more encouraged. It is no wonder, then, that the frequent product of this thought process, the multiple-feature region, is the one that has most attracted those seeking regional patterns and typologies. Although the kinds of multiple-feature regions are naturally multitudinous and there is no universal agreement on the terms applied to them, four general groupings can be recognized: land capability regions, field system regions, farming system regions, and functional regions.

Land-capability regions are the objectives of those seeking a regional classification of one or more aspects of the physical environment with respect to their favorability or unfavorability for agricultural utilization. The more topically inclined agriculturists and pedologists have been most interested in such regionalizations, although geographers, especially the more physically oriented, have also contributed. Potential crop growth has been regionalized in terms of a worldwide climatic system by Papadakis, Thornthwaite, and Bennett, and Nuttonson is approaching this stage with his regularly appearing regional "climatic analogs." Some investigators have tried to improve the reliability of their evaluations by incorporating additional environmental elements into their regional classifications. Hollstein also takes the suitability of the terrain into account when he multiplies the potential caloric production of the principal grain crop in each region by the percentage of the area which is arable. Visher widens the evaluation base considerably further when he lists climate and weather, soils, and accessibility of market as his criteria for constructing a worldwide regional pattern of agricultural potentialities.[6]

Paradoxically, the desire for a regionalization of agricultural potential based on the capabilities of the total physical environment has led many to focus largely on soil conditions. The emphasis is of course natural, for soil is the environmental agent most immediately related to growing

[6] J. Papadakis, *Climates of the World and Their Agricultural Potentials* (Buenos Aires: Privately published, 1960); C. W. Thornthwaite, "An Approach toward a Rational Classification of Climate," *Geographical Review*, Vol. 38 (1948), 55–94; M. K. Bennett, "A World Map of Foodcrop Climates," *Food Research Institute Studies*, Vol. 1 (1960), 285–95; M. Y. Nuttonson, "Crops and Weather," *Landscape*, Vol. 12 (1962), 9–11; W. Hollstein, *Eine Bonitierung der Erde auf landwirtschaftlicher und bodenkundlicher Grundlage*, Petermanns Mitteilungen, Ergänzungsheft 234 (Gotha: Haack, 1937), pp. 17 f.; S. S. Visher, "Comparative Agricultural Potentials of the World's Regions," *Economic Geography*, Vol. 31 (1955), 82–86.

plants and it also is the combined product of climatic, biotic, and geomorphic processes. The stimulus for this view in the United States came in good part from Veatch's proposal, in 1930, for a four-level classificatory hierarchy of "natural land divisions" or "land types." The broadest divisions of land were based essentially on climate—arctic, temperate, and tropic—but the next three, and more significant, levels comprised land types that were homogeneous in topographic, vegetation, and soil conditions. Two of these subtypes were nearly equivalent to the major soil families and soil types. Much of this approach can now be seen in many of the numerous land-evaluation programs being carried out by the state and federal governments, as expressed in the maps of "land capability" and "land resource" regions being produced by the U.S. Soil Conservation Service and cooperating state agencies. Much the same emphasis on soils occurs in the now rapidly accelerating programs in the Soviet Union designed for a more detailed and comprehensive mapping of its land capability regions or "natural-economic zones." [7]

Many researchers, particularly geographers, have strongly criticized the assumption that often accompanies the formation of systems of land capability regions, namely that such units and agricultural regions are generally one and the same thing. Natural units and regions of similar farming types do not as a rule coincide, however, and not uncommonly violently contrast with each other—as Siemens has dramatically shown in his study of farming patterns in a highly variegated physical area in West Germany. Geographers in the Soviet Union have also emphasized the fallibility of environmental evaluations made within the framework of present technical and economic levels, and on this basis deny the validity of any concrete system of areal capability units designed with particular land uses in mind. Only land capability regions having general biogeographic significance are accepted. Even more serious to these men is that the substitution of one region—natural or agricultural—for the other destroys the principal value of both for land assessment by making it impossible to compare them.[8]

Potentialities are replaced by actualities in the study of regions constructed in terms of *field systems,* for they reflect the part of the farming system that is clearest in the landscape, the rotation of uses over the fields. Agricultural economists, however, have traditionally been interested more in the functional than in the spatial patterns of field rotation,

[7] J. O. Veatch, "Natural Geographic Divisions of Land," *Papers of the Michigan Academy of Science, Arts and Letters,* Vol. 16 (1930), 418 f.; M. E. Austin, *Land Resource Regions and Major Land Resource Areas of the United States,* Dept. of Agriculture Handbook No. 296 (Washington, D.C.: U.S. Govt. Printing Office, 1965); N. A. Gvozdetskiy, "Physical-Geographic Regionalization for Agricultural Purposes," *Soviet Geography: Review and Translation,* Vol. 5, No. 7 (1964), 3–10; A. N. Rakitnikov and V. G. Kryuchkov, "Agricultural Regionalization," *ibid.,* Vol. 7, No. 5 (1966), 48–53.

[8] G. von Siemens, "Zur agrargeographischen Landschaftsgliederung: Am Beispiel des südlichen Bergischen Landes erläutert," *Erdkunde,* Vol. 3 (1949), 132–43; A. N. Rakitnikov and I. F. Mukomel, "Agricultural Regionalization," *Soviet Geography: Review and Translation,* Vol. 5, No. 9 (1964), 31; Rakitnikov and Kryuchkov, *op. cit.,* 53, 57.

and geographers have thought of the process only as one, frequently minor, part of the agricultural area. Yet enough representatives of both groups have been interested enough to propose several schemes of regionalization. Before transportation and farm improvements loosened the bond of distance and promoted greater specialization in many areas, farm economists were particularly impressed with the zonal patterns of rotation systems near market centers. Today, the accent is on distributional groupings based on such elements as the length of the growing season, altitude, soil conditions, and farm size. Andreae has accomplished the prodigious task of summarizing, for Western and Central Europe, the spatial reaction of systems of rotation to the combined effect of these physical and cultural elements. The bases for his regional system of two groups of five regions each—"grazing economies" (*Feldgraswirtschaften*) and "cropping economies" (*Felderwirtschaften*)—are the percentage of cropland in fodder crops, the number of years in a rotation fields are kept in fodder crops, and the number of years in a rotation fields are kept in crops other than cereals.[9]

Among the geographers, Otremba believes that sufficient information about field systems is now available so that at least a coarse worldwide regional scheme can be presented. Five principal "land use systems" (*Bodennutzungssyteme*) are offered, ranked by complexity: shifting of pasture, as by the nomads; shifting of fields, as by shifting cultivators; permanent cultivation of a field to the same crop; rotation systems for grazing and cropping economies; and rotation systems for combined grazing and cropping economies. Although the major areas of each field system and its variations are noted, no map is given, presumably because of the many highly complex distribution patterns and the frequently only tentative data. The difficulty of determining the boundaries of the different field combinations, even in localized areas, is shown by the almost-confusing variety of regions mapped by Weaver for the American Midwest. And even where field combinations are more static, as in areas of mixed orchard types, regionalization can be troublesome because of the lack of statistics about acreage. Both Olmstead and Krueger recount this problem in their efforts to outline fruit regions in the United States and Canada.[10]

Regional typologies of the total farming system, and not just that part that is reflected in rotation of fields, have, on the other hand, received considerable attention from both agricultural economists and geographers. The search has entailed constant effort to develop and refine the methods of defining regions of farming systems by uniform criteria.

[9] B. Andreae, *Betriebsformen in der Landwirtschaft* (Stuttgart: Ulmer, 1964), pp. 84–120.

[10] Otremba, *op. cit.*, pp. 158–74; J. C. Weaver, "Crop-Combination Regions in the Middle West," *Geographical Review*, Vol. 44 (1954), Pl. 2, facing p. 182; *idem*, L. P. Hoag, and B. L. Fenton, "Livestock Units and Combination Regions in the Middle West," *Economic Geography*, Vol. 32 (1956), Fig. 14, facing p. 258; C. W. Olmstead, "American Orchard and Vineyard Regions," *ibid.*, Vol. 32 (1956), 195; R. R. Krueger, "The Geography of the Orchard Industry of Canada," *Geographical Bulletin*, Vol. 7 (1965), 30.

Engelbrecht took the first important step in 1883, when he applied the ratio technique to absolute farm data (see Chap. 2). The results of these and later computations were his "agricultural zones." Many scholars, particularly geographers, have criticized these zones as too narrow in scope, since they are characterized by only one crop; what is often overlooked is that by using ratios to determine the importance of the key crop in relation to other crops, Engelbrecht was also providing at least an elementary understanding of the farming structure, something that absolute data were much less able to do.[11]

The next important steps were taken by two Russians, Tschelinzeff and Studensky. Like others, they recognized that Engelbrecht's method only allowed one to compare the individual branches of production within a farming system. A farming system, however, is a complex of correlations which can only be fully perceived and made comparable with those of other systems by converting the data to a single over-all value. This Tschelinzeff did in 1918, when he established farming system regions by applying six kinds of ratios to agricultural census and railroad freight data about the principal provinces of European Russia. Individual rankings of the provinces for each of the ratios were then averaged in order to arrive at an over-all intensity value for each province.[12] Yet weaknesses remained. The single intensity value was still only a relative one, for it was based solely on the performance of one farming system in relation to another. Furthermore, it was a value derived from a comparison of ratios that had nothing to do with monetary values, so that the study, though it presented well the technical aspects of the farm structure, did not reveal its more economic aspects. Moreover, the ratios were arbitrarily chosen and thus were underlain by an intuitive assessment, so that in this computation one unknown was simply expressed by another unknown.

To overcome these drawbacks, Studensky proposed another unit for measuring the intensity of farming systems: gross income. This he applied to agricultural census data about European Russia, averaged over the 1911–1915 period. The gross income of each of the major provinces was computed and then related to a *desyatina* (2.7 acres) of agricultural land; finally, the distribution of gross income among the individual agricultural enterprises was determined. Thus Studensky was able to produce a pattern of agricultural regions through a much more objective and uniform measurement of all of the principal elements of the farm structure. In the same study, he also applied the gross income index to the 1920 agricultural census statistics for the United States, arriving at a pattern of nine state-groups.[13]

[11] Engelbrecht, "Der Standort."

[12] A. Tschelinzeff, *Zustand und Entwicklung der russischen Landwirtschaft nach den Daten der landwirtschaftlichen Betriebszählung von 1916 und der Eisenbahn-Güterverkehrsstatistik* (in Russ.; Kharkov, 1918), cited by G. Studensky, "Die Grundideen und Methoden der landwirtschaftlichen Geographie," *Weltwirtschaftliches Archiv,* Vol. 25 (1927), pp. 185 f.

[13] Studensky, *op. cit.,* pp. 187–97.

Gross income and its distribution among the various income sources became a favorite index among American agricultural economists and geographers. It was one of the main tools used by Baker in his series of detailed regional studies of North American agriculture, the first of which actually appeared one year before Studensky's 1927 study. The most impressive regionalization of American farming systems, however, was completed in 1933 by Elliott and his co-workers in the federal Bureau of Agricultural Economics, using 1930 census data. Twelve major "type-of-farming" regions and one hundred subregions were outlined. Although Elliott provided no over-all ranking of the agricultural intensity of the regions, as Studensky had, his regionalization was much more detailed. Criteria other than the allocation of income were also taken into consideration in the areal differentiation. In 1950 a revised and even more detailed version of this system of agricultural regions was completed. It comprised a three-step hierarchy: 9 "major agricultural regions," 61 "subregions," and 165 "generalized type-of-farming areas." [14]

It is still not feasible, however, to use gross income as a basis for agricultural regionalization in most of the world. Most farmers still sell only a part of their produce, and the statistics about much of what is sold are not readily available. Moreover, farming systems in many areas are comprised of a bewildering variety of crop rotations and animal-crop combinations, the full differences among which would not be revealed by income differences alone. All of this has spurred the quest for a more comprehensive index. For a long time, the most common method of deriving regions on a nonmonetary basis was to determine the relative shares of the agricultural land devoted to the individual crops. In 1930, Brinkmann took note of the fact that areal proportions do not always coincide with the relative economic importance of the various crops in the farming system, and recommended the multiplication of the share of agricultural land in each crop by a weighted index number, or "crop weight" (*Anbaugewicht*). The formula has been used frequently since then, the most detailed application being made by Busch in his delineation of the "agricultural zones" of Central Europe.[15]

A related method has served as a basis for two proposals for an even fuller characterization of the farming system region, by Coppock and by Andreae, both in 1964. Both extended the formula to livestock, for originally it represented them only somewhat indirectly, through fodder crops and pasture. The first move in this extension was the reduction of the livestock types to a common measurable unit, Coppock using feed requirements and Andreae, the density of the livestock population.

[14] O. E. Baker, "Agricultural Regions of North America: Part I. The Basis of Classification," *Economic Geography,* Vol. 2 (1926), 459–93; F. F. Elliott, *Types of Farming in the United States* (Washington, D.C.: U.S. Govt. Printing Office, 1933); *idem, Generalized Types of Farming in the United States,* Dept. of Agriculture Information Bulletin No. 3 (Washington, D.C.: U.S. Govt. Printing Office, 1950).

[15] T. Brinkmann, "Entwicklungslinien und Entwicklungsmöglichkeiten der landwirtschaftlichen Erzeugung Argentiniens," *Berichte über Landwirtschaft,* Vol. 13 (1930), pp. 569 ff.; E. Busch, *Die Landbauzonen im deutschen Lebensraum* (Stuttgart: Ulmer, 1936).

The crop and livestock units were converted to a common unit of measurement by multiplying them by the labor requirements of each unit. Coppock used the "man-day"; Andreae, the "labor hour" (*Arbeitskraftstunde*). The labor indices were then allocated to the appropriate agricultural enterprises, after which it was possible to identify the combinations of enterprises. By these means, Coppock developed the "enterprise-combination" regions of England and Wales, and Andreae set up the "farming system" regions of Western and Central Europe.[16]

The success of these two regionalization methods brings much closer the time when farming system regions, at least those based on labor demands, will be mapped in detail for large areas. Andreae has already specifically demonstrated the practicability of a worldwide classification scheme of this kind, having applied his technique to at least several farming enterprises in each of the major world climatic regions.[17] Because of the larger area he considered, and perhaps also because he did not use a computer, Andreae found himself forced to generalize his regional criteria more than Coppock.

Although they made major strides in the techniques of regionalization, the methods Andreae and Coppock used are not necessarily more objective than those based on income sources. Arbitrary judgments were involved, as shown in the different criteria the two authors used in converting livestock types to a common unit and in the dissimilar labor values the two assigned to the individual crop and livestock types. Both methods also suffer from the problem of all regionalization procedures that use unit-area data instead of individual farm records: the distortion and obscuring of farming systems either because of the originally insufficient breakdown of census statistics for the reporting area or because of the failure of the reporting-area boundaries to coincide with those of the farming systems. Yet one can also be certain that problems like these will receive more attention as interest in the more extensive regionalization of farming systems increases.

Schemes are also being increasingly sought for functional regions, the last of the four general groups of agricultural regions based on multiple features. Interest in them is newer, mostly because the increase in the number and strength of economic ties between different areas has been especially great since World War II. The functioning of these ties is naturally of fundamental concern to most agricultural regions, but attention has been focused on functions only insofar as they form a part of the regional complex, not as they reflect the more extensive and less space-bound interaction of supply and demand. It is the spatial expression of this reflection that constitutes the functional agricultural regions.

The broadest view of the functional agricultural regions involves the arrangement of states with respect to the degree of balance obtained

[16] J. T. Coppock, "Crop, Livestock, and Enterprise Combinations in England and Wales," *Economic Geography,* Vol. 40 (1964), 65–81; Andreae, *op. cit.,* pp. 61-75.

[17] *Ibid.,* pp. 166–318.

between supply and demand. Not much has been done on this subject, particularly on a worldwide basis. One reason has been the many criteria demanding consideration: the relation of exports and imports to production, the relation of exports and imports computed on a per capita basis, the share of the agricultural population, and the orientation of production—ranging from monoculture to extremely complex mixed farming—are only the more important ones. The necessary data are often lacking or unreliable, although the increasing number of statistical reports from national and international agencies, such as the Food and Agriculture Organization, are beginning to provide a basis for more and surer worldwide regional comparisons. Otremba has briefly discussed a typology of "agricultural-surplus," "agricultural-deficiency," and "agricultural-autarkic" areas, and both he and Gottmann indirectly note such areas in their studies of the distribution of individual markets for the major agricultural commodities. The possibility of employing a common unit for measuring the agricultural sufficiency of a country, as has been done for farming system regions, has been hinted at by Hoffmann. Although he does not specifically rank sufficiency, he comes close by dividing groups of countries ("great regions") into "high-income" and "low-income" categories according to their shares of total world agricultural output, which he expresses in "value added," or the gross value of output minus purchase costs.[18]

More regionalization of supply-and-demand functions has been attempted for centrally oriented agricultural areas, as can be seen in the variety of approaches and the expression of some dissenting opinions. Some workers have concerned themselves more with the supply role of the agricultural area, others with the demand function, and within these two orientations different paths have also been taken. Thus, among those who put the supply aspect in the foreground, Eyre has set up a hierarchy of regions (Japanese prefectures) on the basis of volume of fresh food supplied a single market (Tokyo); Durand has simply outlined a pattern of regions, but with relation to the locations of several markets (Northeastern U.S. milksheds); and Gregor has combined the two approaches in his hierarchization of regions (California county groupings) on the strength of their production of fresh produce for two markets, Los Angeles and the San Francisco Bay Area.[19]

Among those interested more in the demand character of functional agricultural regions, the divergences in outlook exist primarily in the methods favored for defining the extent of influence of rural service

[18] Otremba, *op. cit.*, pp. 58 f.; *idem, Allegemeine Geographie des Welthandels und des Weltverkehrs* (Stuttgart: Franckh, 1957), pp. 200–25; J. Gottmann, *Les marchés des matières premières* (Paris: Armand Colin, 1957), pp. 81–212; L. A. Hoffman, *Economic Geography* (New York: Ronald Press, 1965), p. 97.

[19] J. D. Eyre, "Sources of Tokyo's Fresh Food Supply," *Geographical Review*, Vol. 49 (1959), 455–74; L. Durand, Jr., "The Major Milksheds of the Northeastern Quarter of the United States," *Economic Geography*, Vol. 40 (1964), 9–33; H. F. Gregor, "The Local-Supply Agriculture of California," *Annals of the Association of American Geographers*, Vol. 47 (1957), 267–76.

centers. Brush found that traffic divides, shown in the pattern of daily traffic movement toward the center and other centers, were the best means of determining not only the pattern of service areas in southwestern Wisconsin, but also a two-level hierarchy for them. Schroeder came to the opposite conclusion in his examination of the influence patterns of service centers in the lower Rio Grande Valley of Texas, stating that motorization caused so many overlaps that it was impossible to clearly bound the service areas. Bracey is more optimistic, like Brush, but differs with him by rejecting traffic as a delimiter of service areas in favor of the number of village residents that visit the service centers. This was a logical conclusion for him in his study of service areas in southern England, where the larger number of service centers and greater mix of urban and rural functions made a sharper distinction of rural uses necessary. Like Brush, Bracey found it possible to delimit two orders of service areas.[20]

The Total Agricultural Region

If agricultural regions can be viewed as spatial units consisting of single features or certain combinations of several features, then they can be no less logically seen in their totality. Geographers in particular have been interested in the over-all agricultural region, although they by no means completely agree on its definition or the importance of its study to agricultural geography.

Physical, economic, and cultural biases all stand out in the differing conceptions of the total agricultural region, despite reference by all authors to a wide range of regional characteristics. Baker's definition typifies the physical emphasis. For him, the region was an "area of land characterized by homogeneity of agricultural conditions, especially crops grown." Further, it was "usually determined principally by climatic conditions." Subdivisions of the region were the results of "differences in land reliefs, or in slope, and in soils," which might cause variations in the proportion of the land used for crops, pasture, or forest, or in the relative importance of the crops. Economic principles were important also, but only in their capacity to "strengthen the influence of the physical factors" to join with them in determining the boundaries of the region.[21]

Economic factors come to the fore in the definitions of Whittlesey, Hartshorne, and Dicken, and emphasis shifts from the crop to the operational structure of the farm. For Hartshorne and Dicken, this structure was represented by the relative importance of different crops and livestock products on the farm. Whittlesey agreed, putting the kind of crop and livestock association at the top of a list of five "functioning forms"

[20] J. E. Brush and H. E. Bracey, "Rural Service Centers in Southwestern Wisconsin and Southern England," *Geographical Review,* Vol. 45 (1955), 562 f.; K. Schroeder, *Agrarlandschaftsstudien in südlichsten Texas* (Frankfurt am Main: Kramer, 1962), pp. 162 f.

[21] Baker, *op. cit.,* pp. 468–71, 479.

hat "appear to dominate every type of agriculture." The other "forms" were the methods of production, the intensity of land use and production, the disposal of the products (whether for subsistence or sale), and the ensemble of structures used in the farming operation.[22]

A broader and more cultural concept of the agricultural region, however, is espoused by writers like Waibel, Cholley, and Carol. Waibel viewed the "agricultural formation" (see Chap. 1) as a product not only of natural and economic forces, but of the "entire range of human forces as they are reflected in number and distribution over the earth surface and in social, economic, cultural, and above all intellectual, differentiation." For Cholley, such a region represented a "profound solidarity" between all the elements that are brought together in the "agrarian structure" (*structure agraire*): physical and biological, human, political, and economic. Carol likewise emphasized the "coexistence" and "interlacing" of the "constituting components" of the "agricultural landscape" (*Agrarlandschaft*), in this case fourteen of them, six of which were physical (subsoil, relief, climate, water, soil, natural vegetation), two "cultural" (cultural vegetation, cultural structures), and six "functional" (rural population, cultural and technological stage, farming operation, market, organization for providing the rural population with economic and cultural goods, commerce).[23]

All three of these emphases in the definition of the total agricultural region are well represented in the literature, albeit not so often in as isolated a form or in quite the same terms as posed by the authors just cited. A traditional practice among many geographers engaged in the study of a single region has been to combine something of both the physical and economic approaches: the characterization of the region by its most important crop or crops, and the way those commodities fit into the economic system. Most such characterizations have actually been a byproduct of research on the geographic aspects of the commodity, although where the commodity has been particularly influential in the economy of an area, the result has often been a surprisingly extensive view of the region. Colby's early study of the raisin industry of the San Joaquin, for example, is still regarded as one of the magistral regional studies in American agricultural geography. A more deliberate emphasis on the commodity–region symbiosis is well exemplified by Durand's

[22] R. Hartshorne and S. N. Dicken, "A Classification of the Agricultural Regions of Europe and North America on a Uniform Statistical Basis," *Annals of the Association of American Geographers*, Vol. 25 (1935), 99–120. Despite the close relationship of their method to Engelbrecht's, the two authors also considered landscape appearance. See, e.g., R. Hartshorne, *The Nature of Geography* (Lancaster, Penn.: Association of American Geographers, 1939), p. 345; D. Whittlesey, "Major Agricultural Regions of the Earth," *Annals of the Association of American Geographers*, Vol. 26 (1936), 209.

[23] L. Waibel, *Probleme der Landwirtschaftsgeographie* (Breslau: Hirt, 1933), p. 8; A. Cholley, "Problèmes de structure agraire et d'économie rurale," *Annales de géographie*, Vol. 60 (1946), 83, 86; H. Carol, "Das agrargeographische Betrachtungssystem; ein Beitrag zur landschaftskundlichen Methodik, dargelegt am Beispiel der Karru in Südafrika," *Geographica Helvetica*, Vol. 7 (1952), 20, 22.

studies of the "cheese regions" in Wisconsin and Illinois and the "burley tobacco region" in Tennessee.[24]

Writings about the type of agricultural region formulated by geographers like Waibel have been generally more straightforward, if for no other reason than the far more comprehensive view demanded of the researcher. This is evident in the voluminous monographs by Brunet on the "agrarian structure" east of Paris and by Phlipponneau on "rural life" on the margins of that city, two of the best examples of the kind of work that has become a tradition among French geographers. Waibel, however, a German, must be given credit for one of the earliest attempts to discern this broader and more culturally oriented agricultural region, in his 1933 work on the Sierra Madre de Chiapas, which, incidentally, preceded by only a year the first major effort of this kind in the United States, by Gibson on the "land economy" of the central Pennyroyal in Kentucky. Other examples of this regional approach include such excellent works as Bartz's, on the "polder landscape" of the Sacramento–San Joaquin Delta in California, Dobby's, on the Kelantan Delta, and Robinson and Burley's, on the Maitland Flats in New South Wales. To this list should also probably be added the several plantation studies by Prunty, which, though concentrating upon one farm type that is frequently dispersed among others, thoroughly pictures the geographic reality of individual plantations.[25]

The more comprehensive view implicit in regionalization based on total agriculture has also been a primary reason for the scarcity of relevant worldwide classification schemes. In fact, only two classifications—one published by Whittlesey in 1936 and another by Hahn in 1892—have gained widespread acceptance, and only Whittlesey's purports to illustrate one of the three versions of the total region. A third worldwide classification is based on Baker's approach, but only in full detail for

[24] C. C. Colby, "The California Raisin Industry: A Study in Geographic Interpretation," *Annals of the Association of American Geographers,* Vol. 14 (1924), 50–108; L. Durand, Jr., "Cheese Region of Southeastern Wisconsin," *Economic Geography,* Vol. 15 (1939), 283–92; *idem,* "Cheese Region of Northwestern Illinois," *ibid.,* Vol. 22 (1946), 24–37; *idem* and E. T. Bird, "The Burley Tobacco Region of the Mountain South," *ibid.,* Vol. 26 (1950), 274–300.

[25] P. Brunet, *Structure agraire et économie rurale des plateaux tertiares entre la Seine et l'Oise* (Caen: Caron, 1960); M. Phlipponneau, *La vie rurale de la banlieue parisienne* (Paris: Armand Colin, 1956); L. Waibel, "Die Sierra Madre de Chapas," *Mitteilungen der Geographischen Gesellschaft zu Hamburg,* Vol. 43 (1933), 102–44; J. S. Gibson, "Land Economy of Warren County, Kentucky," *Economic Geography,* Vol. 10 (1934), 74–98, 200–216, 268–87; F. Bartz, "Die Polderlandschaft des Deltas des Sacramento-San Joaquin: Das Holland Kaliforniens," *Erdkunde,* Vol. 6 (1952), 247–62; E. H. G. Dobby, "The Kelantan Delta," *Geographical Review,* Vol. 41 (1951), 226–55; K. W. Robinson and T. M. Burley, "Flood-Plain Farming on the Maitland Flats," *Economic Geography,* Vol. 38 (1962), 234–50.

Of Prunty's work, see particularly "The Woodland Plantation as a Contemporary Occupance Type in the South," *Geographical Review,* Vol. 53 (1963), 1–21; and "Deltapine: Field Laboratory for the Neoplantation Occupance Type," in *Festschrift: Clarence F. Jones,* ed. M. Prunty, Jr. (Evanston, Ill.: Northwestern Univ., Dept. of Geography, 1962), pp. 151–72.

he United States and Canada. The other continents are each treated by a different author with his own regional system.[26]

The widespread acceptance by geographers of Whittlesey's and Hahn's classifications has been long-term. Both of the maps, with occasional modifications and generalizations, have appeared in consecutive editions of many atlases and economic geography texts. Their popularity, however, is sharply divided between scholars of North America and Britain and those of continental Europe. The first group favors the strongly economically oriented regional plan of Whittlesey. Using the five "functioning forms" already noted, as well as field observations, Whittlesey constructed thirteen "major agricultural regions of the earth": (1) nomadic herding; (2) livestock ranching; (3) shifting cultivation; (4) rudimental sedentary tillage; (5) intensive subsistence tillage, rice dominant; (6) intensive subsistence tillage without wet rice; (7) commercial plantation crop tillage; (8) Mediterranean agriculture; (9) commercial grain farming; (10) commercial livestock and crop farming; (11) subsistence crop and livestock farming; (12) commercial dairy farming; and (13) specialized horticulture. Hahn, in contrast, stressed more the cultural outlook, as is evident in most of the titles he selected for his six regions or "economic forms" (*Wirtschaftsformen*): (1) hunting and fishing; (2) hoe culture; (3) plantation culture; (4) plow culture; (5) animal husbandry; and (6) garden culture. Thus he emphasized more those things that concern rural man, rather than the agricultural entrepreneur: kinds of work tools and the way they are used; sizes and shapes of fields in relation to those tools and work methods; assignments of work among the sexes and age groups; association of work patterns and different products with traditional ceremonies; and the like.[27]

Neither regional classification, however, despite its wide following, has escaped criticism. Complaints have become particularly sharp as the agricultural character of widespread areas continues to change and thus make the patterns obsolete.[28] Such changes, however, seem to some to necessitate only refinements of the original pattern: Symons, for example, suggests for the Whittlesey classification additional regional types and combinations of older types because of the increase in commercialized agriculture. But Otremba, in his critique of Hahn's regional cate-

[26] Whittlesey, *op. cit.*, pp. 149–240; E. Hahn, "Die Wirtschaftsformen der Erde," *Petermanns Mitteilungen*, Vol. 38 (1892), 8–12.

The worldwide classification is a world-region series that appeared in the periodical, *Economic Geography*. Authors and the continents they investigated were Baker (North America), Vols. 2–9 (1926–33); O. Jonasson (Europe), Vols. 1–2 (1925–26); C. F. Jones (South America), Vols. 4–6 (1928–30); H. L. Shantz (Africa), Vols. 16–19 (1940–43); G. Taylor (Australia), Vol. 6 (1930); and S. Van Valkenburg (Asia), Vols. 7–12 (1931–36).

[27] A succinct description of Hahn's economic forms is given by R. Lütgens in his *Die geographischen Grundlagen und Probleme des Wirtschaftlebens* (Stuttgart: Franckh, 1950), pp. 173–77. Some of the characteristics of these forms are also described in H. F. Gregor, *Environment and Economic Life* (Princeton, N.J.: Van Nostrand, 1963), pp. 287–306.

[28] A problem well summarized in D. Grigg, "The Agricultural Regions of the World: Review and Reflections," *Economic Geography*, Vol. 45 (1969), 113 f., 117–20.

gories, calls for their elimination because of increasing intermixtures of farming types. He would substitute four "economic forms," based on production goals and social organization, criteria that George also later used for his four "agricultural forms" and Enyedi for his "agricultural types." A milder view is taken by Gregor, who believes it is possible to recognize six "combination forms" in addition to Hahn's original designations. But areal delimitation has seemingly proved even more of a problem than classification, for only a few of the more recent proposals for alterations of Hahn's and Whittlesey's regional patterns have been accompanied by maps. One of these is the revision by Trewartha, Robinson, and Hammond of Whittlesey's map; another is the author's world map of agricultural regions based on Hahn's classification (Figs. 6 and 7).[29]

The critics of Whittlesey's classification nevertheless can be at least more satisfied than most critics of Hahn's scheme, if for no other reason than the differing conceptions of the total agricultural region. Most of those who view this areal unit as primarily an economic one still find in many of Whittlesey's criteria its best expression to date. But by those who view the total agricultural region as more broadly based, improvements in Hahn's system can only be considered progress in the worldwide regionalization of just one, albeit critical, facet of the region, the cultural. Whether a worldwide regional system can actually be accomplished for this kind of region will be known only after much more investigation. Typologies have already been suggested by Obst and Carol, however, and Credner has proposed the use of a fractional code that would enable uniform mapping of such regions at large and medium scales.[30] Regionalization of this type has also been applied to such widely separated areas as California and Germany. The pattern of "rural regions" established for the United States by the sociologists like Taylor is extremely close to this concept of the total agricultural region.[31]

Criticism of schemes of regionalization is not confined to those interested in refining regional patterns or discovering new ones. Many geographers seriously question their value, either wholly or in part. Anuchin sees no validity in regionalization as an end in itself, though he believes it an important tool in geographic research. McCarty and Lind-

[29] L. Symons, *Agricultural Geography* (New York: Praeger, 1967), pp. 170–76; Otremba, *Allgemeine Agrar- und Industriegeographie,* pp. 213–19; P. George, *Précis de géographie rurale* (Paris: Presses univ. de France, 1963), pp. 195–289; G. Enyedi, *The Agriculture of the World (A study in agricultural geography),* Abstracts, No. 9 (Budapest: Hungarian Academy of Sciences, Institute of Geography, 1967); G. T. Trewartha, A. H. Robinson, and E. H. Hammond, *Elements of Geography,* 5th ed. (New York: McGraw-Hill, 1967), Appendix 8; Gregor, *Environment and Economic Life,* back end paper.

[30] E. Obst, "Das Problem der allgemeinen Geographie," *Verhandlungen des deutschen Geographentages,* Vol. 2 (1950), 45 f.; Carol, *op. cit.,* p. 23; W. Credner, "Mitteilungen der landwirtschaftsgeographischen Arbeitsgemeinschaft 1943," unpublished manuscript cited by Otremba, *Allgemeine Agrar- und Industriegeographie,* pp. 55 f.

[31] G. Pfeifer, *Die räumliche Gliederung der Landwirtschaft im nördlichen Kalifornien* (Leipzig: Hirt, 1936), pp. 46–83, 139–292; E. Otremba, *Die deutsche Agrarlandschaft,* 2d ed. (Wiesbaden: Steiner, 1961); C. C. Taylor *et al., Rural Life in the United States* (New York: Knopf, 1949), pp. 329–491.

berg go even further, seeing regionalization as principally a descriptive device and rarely as a tool in analysis. This judgment they base on the fact that many economic functions that give character to a region are also parts of a much wider economic process that knows no inner boundaries. The question of just how long these functions can develop, however, without evoking regional expressions in function and landscape has been raised by Otremba.[32] Further, as we saw in the opening chapter, the view that the region is not of primary value to geographic research is based on objectives which are incapable of achievement by the traditional regional approach and which are vigorously disputed.

Even those convinced of the efficacy of the region, though, are not unanimously in favor of all attempts at regionalization. One of the most ardent defenders of the regional viewpoint, Hartshorne, is also critical of attempts to set up hierarchies of regions and subregions. He points to the varying importance of the independent factors making up these units as the principal obstacle to any logical hierarchization. Grigg still advocates the approach as a means for making generalizations about the same objects at different levels of abstraction. Many more geographers, however, are of the opinion that the setting up of regional patterns of any kind should proceed more slowly until a firmer foundation of facts can be developed. This is a concern of long standing, most forcefully expressed in the reservations of Dion and Weaver. It has also been shown most recently in the complaints of participants, in a symposium on a worldwide typology of agriculture, over the unavailability and heterogeneity of data.[33] It is also interesting that these cautionary views are the bluntest rejections of the implications of the criticism of regionalization by theoretical geographers, namely, that analysis is more important than description.

Comparative Regional Studies

Another regional approach, and for many the crowning step in regional study, is regional comparison. The principal objective of those using it has been the fuller comprehension of the regions, for "as the personality of man is first recognized in society, so is the individuality of

[32] V. A. Anuchin, "On the Subject of Economic Geography," *Soviet Geography: Review and Translation,* Vol. 2, No. 3 (1961), 34; H. H. McCarty and J. B. Lindberg, *A Preface to Economic Geography* (Englewood Cliffs, N.J.: Prentice-Hall, 1966), pp. 95 f., 103 f.; E. Otremba, "Struktur und Funktion im Wirtschaftsraum," *Festschrift Theodore Kraus: Wirtschafts- und sozialgeographische Themen zur Landeskunde* (Bad Godesberg: Bundesanstalt für Landeskunde und Raumforschung, 1959), pp. 22 f.

[33] R. Hartshorne, *Perspective on the Nature of Geography* (Chicago: Rand McNally, 1959), p. 132; D. Grigg, "The Logic of Regional Systems," *Annals of the Association of American Geographers,* Vol. 55 (1965), 489 f.; R. Dion, "À propos de géographie agraire: Les faits avant les théories," *Annales de géographie,* Vol. 58 (1949), 339–41; J. C. Weaver, "A Design for Research in the Geography of Agriculture," *Professional Geographer,* Vol. 10 (1958), No. 1, 2–8; F. A. Barnes, reporter, "The Geographical Typology of Agriculture," in *Agricultural Geography I.G.U. Symposium,* ed. E. S. Simpson (Liverpool: Dept. of Geography, Univ. of Liverpool, 1965), pp. 59–74.

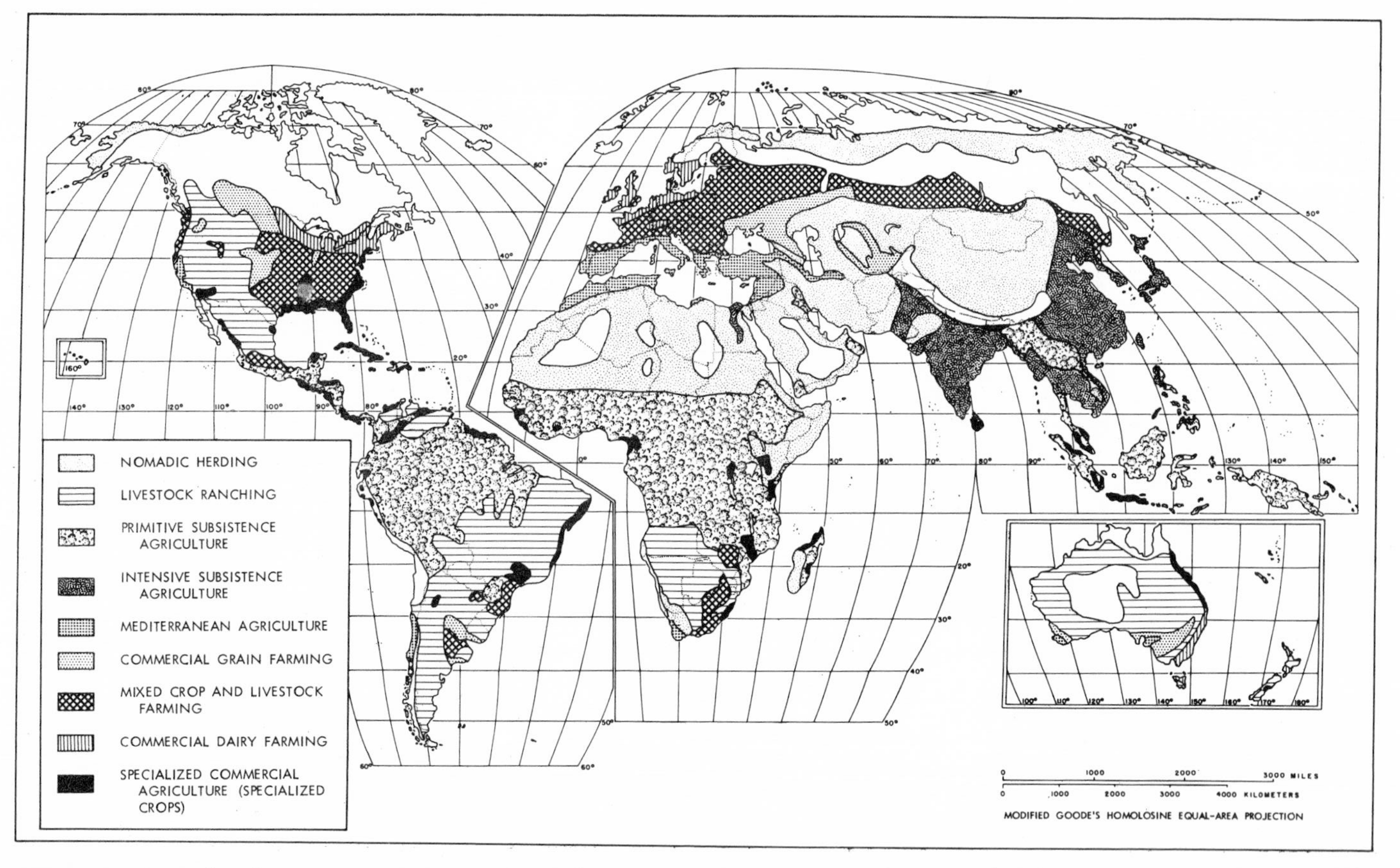

Figure 6. *Agricultural regions, after Trewartha, Robinson, and Hammond. [Adapted from* Elements of Geography. *Copyright 1967 by McGraw-Hill, Inc. By permission of McGraw-Hill Book Company.]*

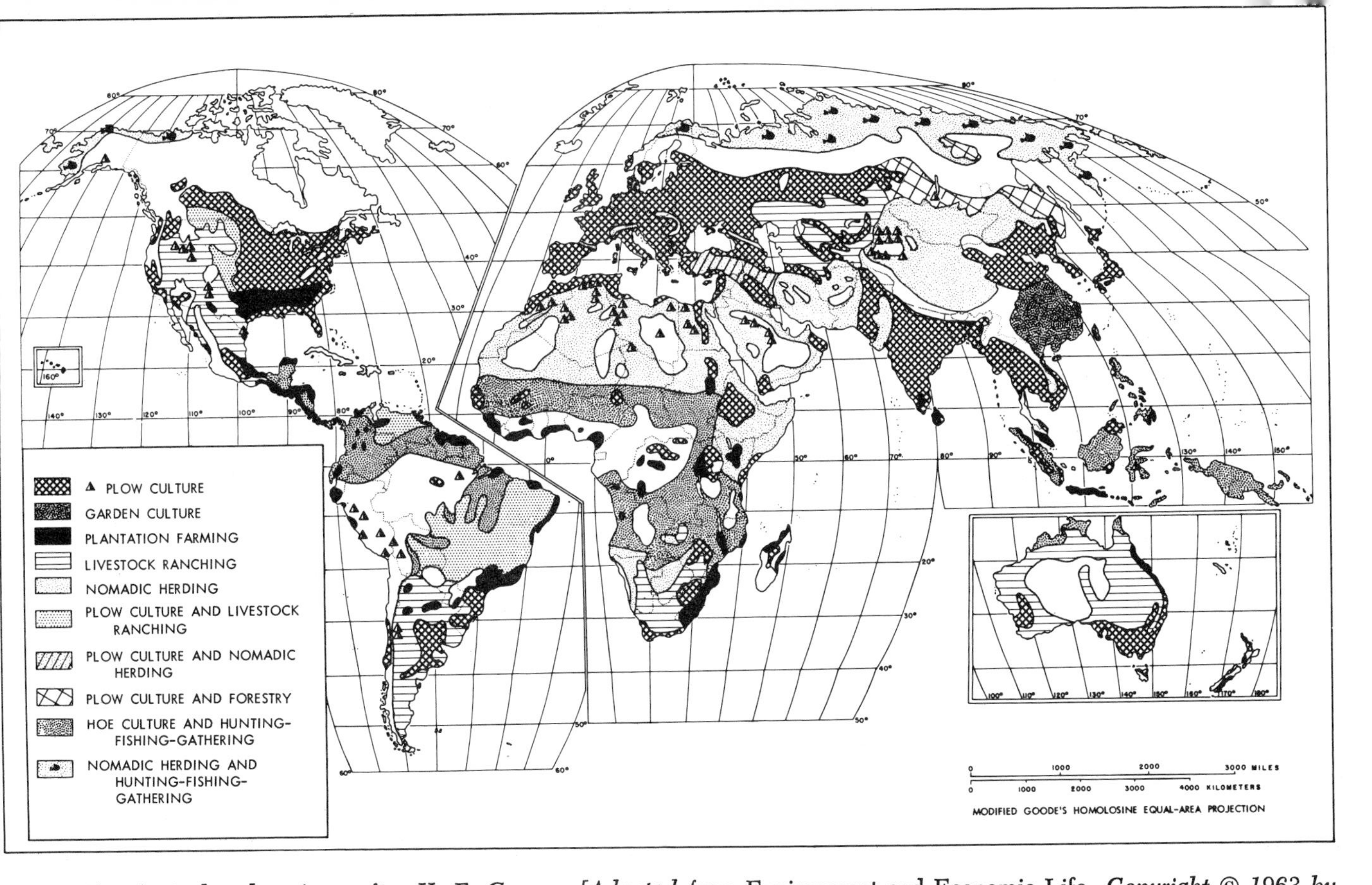

Figure 7. *Agricultural regions, after H. F. Gregor. [Adapted from* Environment and Economic Life. *Copyright © 1963 by D. Van Nostrand Reinhold Company.]*

area first made clear in association with other near and distant areas."[34] The regions compared have usually been similar ones, since it was the similarities between widely separated regions that first attracted the investigator and then encouraged him to search for the contrasts. Similarities also reduce the variables, so that more generalizations can be made. Yet despite its attractions and benefits, the literature on comparative regional studies is a small part of the total writings on regions in agricultural geography. The need for close familiarity with more than one region is undoubtedly one reason for the weak showing.

Those comparative regional studies that have been made nevertheless cover a surprising variety of subjects, particularly the more recent works. By inventorying temperature and moisture of the arable land resources of the Soviet Union and North America (the United States and Canada), Field has shown deep contrasts between two agricultural land bases that otherwise display striking similarities. Similar contradictions have been described by Meinig and Ehlers in their comparisons of colonization, Meinig having compared the American and Australian wheatlands, and Ehlers the Canadian and Finnish boreal forest lands. Lewthwaite accomplished what could be considered the ultimate in regional agricultural geography, a comparison of two total agricultural regions, the American and New Zealand dairy regions, as exemplified by Wisconsin and the Waikato. The total agricultural region in miniature, the farm, has also begun to be studied in more comparative detail, with such results as Siemens finding, in a part of the Rhine hill country, that different farm sizes result in different land uses within physically similar environments, and Gregor finding, in a comparison of California and New Jersey vegetable farms, enough differences between farms of different sizes to suggest that such contrasts might be more prominent regionally than those due to type of specialization.[35]

The traditional question of whether intensive regional comparison can produce universal, as well as unique, generalizations continues to interest those contributing to the literature in agricultural geography. Even more important, some of the recent articles have dealt with regions most resistant to extensive generalization, namely, those areas based more on human than environmental forces. Brush and Bracey, after finding similarities in the distribution patterns of rural service centers and their tributary areas in southwestern Wisconsin and southern England,

[34] E. Otremba, "Das Spiel der Räume," *Geographische Rundschau,* Vol. 13 (1961), p. 133.

[35] N. C. Field, "Environmental Quality and Land Productivity: A Comparison of the Agricultural Land Base of the USSR and North America," *Canadian Geographer,* Vol. 12 (1968), 1–14; D. W. Meinig, "Colonization of Wheatlands: Some Australian and American Comparisons," *Australian Geographer,* Vol. 7 (1959), 205–13; E. Ehlers, "Das boreale Waldland in Finnland und Kanada als Siedlungs- und Wirtschaftsraum," *Geographische Zeitschrift,* Vol. 55 (1967), 280–322; G. R. Lewthwaite, "Wisconsin and the Waikato: A Comparison of Dairy Farming in the United States and New Zealand," *Annals of the Association of American Geographers,* Vol. 54 (1964), 59–87; Siemens, *op. cit.,* pp. 137–43; H. F. Gregor, "Farm Structure in Regional Comparison: California and New Jersey Vegetable Farms," *Economic Geography,* Vol. 45 (1969), 209–25.

feel justified in suggesting that "the spatial hierarchy of central places is related to distance factors that have a dominant influence in areas of low relief and fairly uniform rural population distribution, despite differences in population density, economic functions, and social or political institutions." Spencer and Horvath accept the even greater challenge of drawing universals from a comparison of total agricultural regions, in this case the Corn Belt, the "Philippine coconut landscape," and the "Malayan rubber landscape." These they use to demonstrate their belief in the possibility of identifying six categories of cultural processes operative in the "cultural origin, maturity, and change" of the agricultural region: psychological, political, historical, technologic, economic, and agronomic. Additional support for this emphasis comes from Otremba. After comparing four German agricultural regions, two with similar physical environments but different agricultural landscapes and two with the reverse situation, he finds it logical to erect a regional typology based on the length and character of historical development.[36]

[36] Brush and Bracey, *op. cit.*, p. 568; J. E. Spencer and R. J. Horvath, "How Does an Agricultural Region Originate?" *Annals of the Association of American Geographers,* Vol. 33 (1963), 74–92; Otremba, *Allgemeine Agrar- und Industriegeographie,* 1st ed. (1953), pp. 187–91.

CHAPTER NINE

Regional boundaries

The Agricultural Margins

Delimitation goes hand in hand with regionalization, and for many investigators this objective has been at least as attractive as the characterization of the core of a region. Not surprisingly, those most interested have been those studying the margins of agriculture, where even whole regions reflect the successes and failures of farmers in preserving and advancing the ecumene. The greatest attention has been accorded the dry boundaries, in good part because agricultural advances there have been more numerous and vivid than along the polar and altitude boundaries. By 1932, there were enough data to encourage Sapper to take the most advanced step, the construction of a world map showing the actual limits of irrigation and of livestock raising dependent on artesian wells. By 1965, still more information enabled Highsmith to update and greatly refine the boundaries of irrigated land on his world map. Less work has been done on the potential limits of dry-land agriculture, the best of the recent studies concentrating on the Soviet dry lands, where planning for the advancement of the cropping boundary has been most active. The maps by Ivanova and Fribland, and by Jackson, on the "virgin land" reserves, and by Lewis on just the irrigable lands, have been particularly enlightening.[1]

[1] K. Sapper, "Die Verbreitung der künstlichen Feldbewässerung," *Petermanns Geographische Mitteilungen,* Vol. 78 (1932), 225–31, 295–301, and Tafel 7; R. M. Highsmith, Jr., "Irrigated Lands of the World," *Geographical Review,* Vol. 55 (1965), 382–89 and Pl. 1; E. N. Ivanova and V. M. Fribland, "Osvoenie tselinnykh i zalezhnykh zemel," ["The Assimilation of Virgin and Idle Lands"], *Prireda,* No. 4 (1954), pp. 3–10; W. A. D. Jackson, "Virgin and Idle Lands of Western Siberia and Northern Kazakhstan: A Geographical Appraisal," *Geographical Review,* Vol. 46 (1956), 1–19; R. A. Lewis, "The Irrigation Potential of Soviet Central Asia," *Annals of the Association of American Geographers,* Vol. 52 (1962), 99–114.

Concern with the potential polar limits of agriculture is also lively. Boundaries have been drawn in the Canadian northlands for more than half a century, but none has been universally accepted.[2] The most comprehensive attempt was made by Vanderhill, who has gone beyond the usual postulation of an isotherm and delimited all areas in the Prairie Provinces believed capable of future agricultural settlement. The actual rather than the potential advance of the polar agricultural boundary has been emphasized in the Old World. The rapid poleward expansion of agriculture in the Soviet Union is certainly one reason; also responsible, perhaps, is the simple fact that this stormy push of the farming boundary to the north has drastically reduced the amount of potentially arable land to be studied, a development well documented by Müller-Wille and Buchholz in their cartographic representation of past and present crop boundaries in the Soviet Union.[3]

The extreme fragmentation of the altitudinal boundary of agriculture has long defied attempts to map that line over a sizable area. Most geographers interested in such possibilities derive vertical land use zones by averaging the high points in use in various locations of the particular area. Peattie, however, was an early objector to vertical zoning even of a small area, claiming that the averages derived seldom corresponded with reality. He further asserted that, contrary to the still-common assumption, physical controls are subordinate to economic forces in determining the position of the altitudinal boundary between forest and cropland, and that climate is often less important than the availability of soil in determining the limits of height. Curiously enough, the same conclusion about the lesser importance of climate was later reached by Hambloch in a study that used the approach to altitudinal boundaries that Peattie decried. His was the first major quantitative analysis of the altitudinal limit of the ecumene over the entire earth. However, Hambloch concluded that the principal physical agent was terrain, and not soil.[4]

The problem of reconciling averages with reality is not peculiar to research on altitude. Agricultural margins on the flat are also often fragmented, and writers usually draw the boundary no farther out than the

[2] The attempts are summarized by W. A. Mackintosh, *Prairie Settlement: The Geographical Setting,* Vol. 1 of *Canadian Frontiers of Settlement,* ed. W. A. Mackintosh and W. L. G. Joerg. (Toronto: MacMillan, 1934), pp. 186–200; and by M. K. Bennett, "The Isoline of Ninety Frost-Free Days in Canada," *Economic Geography,* Vol. 35 (1959), 41 f.

[3] B. G. Vanderhill, "The Direction of Settlement in the Prairie Provinces of Canada," *Journal of Geography,* Vol. 58 (1959), p. 326, Fig. 1; W. Müller-Wille, "Europa: Seine Bevölkerung, Energieleistung und Ländergruppen," *Verhandlungen des deutschen Geographentages,* Vol. 34 (1965), 85 f.; E. Buchholz, *Die Waldwirtschaft und Holzindustrie der Sowjetunion* (Munich, 1961), cited by E. Fels, *Der wirtschaftende Mensch als Gestalter der Erde,* 2d ed. (Stuttgart: Franckh, 1967), p. 175.

[4] R. Peattie, "Height Limits of Mountain Economies: A Preliminary Survey of Contributing Factors," *Geographical Review,* Vol. 21 (1931), 415; H. Hambloch, *Der Höhengrenzsaum der Ökumene* (Münster: Institut für Geographie und Länderkunde und Geographischen Kommission für Westfalen, 1966).

edge of the compact mass of farmland. Many have written about agricultural "islands" or "foreposts," but few have studied them extensively and even fewer have mapped their specific extent. Much of this work seems to have been done in South America. Especially searching has been the attention given to the "forest" agricultural boundary in the eastern valleys of the Andes, Monheim, Fifer, and Eidt adding substantially to earlier information. A similar updating of the North American polar farming fringe, in both textual and cartographic detail, can be readily perceived by comparing the earlier materials with those of Vanderhill, Ehlers, and Francis.[5] Another fragmented agricultural boundary, of the oases in the deserts of northwestern Mexico and southeastern California, has been well described and outlined by several recent investigators.[6]

The mapping of agricultural islands on the frontier is also becoming more detailed. The land use complexes of whole townships in the Peace River area have been mapped, for example, as has the distribution of farms and farmsteads in large parts of Ontario and Finland, respectively. A more dynamic dimension has also been provided for this last pattern in a map showing the directions of "attack on woodland and swamp" taken by the farmers as they enlarge the areas of the many farm clusters. Comprehensive mapping of this kind, particularly when done over several areas, promises highly fruitful regional comparisons. Thus Born, after extensive field mapping of agricultural islands along several parts of the arid margins in Sudan, has become convinced that the outermost agricultural boundary is determined not just by rainfall, as generally supposed, but also by the texture of soil. Most of the recent advances of agriculture into the desert have been into areas that have insufficient rainfall but whose soils are capable of receiving and retaining the runoff from moister areas. A comparison of the polar agricultural margins in Finland and Canada by Ehlers, however, has revealed to him marked differences between the farmers' evaluations of soil moisture in the two countries. Finnish farmers consider wetter soils less a hindrance to agriculture if the finer timber of the better-drained land can be preserved. The result is an extremely fragmented agricultural border. Canadian farmers, on the other hand, judge overly moist land fit only for forest, as they prize the better-drained areas much more for agriculture

[5] F. Monheim, *Junge Indianerkolonisation in den Tiefländern Ostboliviens* (Braunschweig: Westermann, 1965); J. V. Fifer, "Bolivia's Pioneer Fringe," *Geographical Review*, Vol. 57 (1967), 2–23; R. C. Eidt, "Japanese Agricultural Colonization: A New Attempt at Land Opening in Argentina," *Economic Geography*, Vol. 44 (1968), 1–20; B. G. Vanderhill, "The Farming Frontier of Western Canada: 1950–1960," *Journal of Geography*, Vol. 61 (1962), 13–20; E. Ehlers, "Landpolitik und Landpotential in den nördlichen kanadischen Prärieprovinzen," *Zeitschrift fur ausländische Landwirtschaft*, Vol. 50 (1966), 42–55; and K. E. Francis, "Outpost Agriculture: The Case of Alaska," *Geographical Review*, Vol. 57 (1967), 496–505.

[6] C. L. Dozier, "Mexico's Transformed Northwest," *ibid.*, Vol. 53 (1963), 548–71. The most comprehensive boundary delineation in southeastern California is H. F. Gregor's "An Evaluation of Oasis Agriculture," *Yearbook of the Association of Pacific Coast Geographers*, Vol. 21 (1959), 39–51.

than forestry. Their intrusions into the forest are therefore much more compact.[7]

The area speckled by agricultural outliers has commonly been accepted as one of comparatively low farming intensity, an agricultural fringe or "sub-ecumene" sandwiched between the more or less compact farmlands, or "full ecumene," and the land that is completely devoid of farmland and people, or "anecumene." The first representations of this area as a transition zone were made on a generalized world scale, beginning with Bowman's map of "pioneer belts" in 1931 and continuing through Pelzer's (1937) and Jaeger's (1946) boundary proposals. Although each map shows basic disagreements with the others and the boundary has been altered in many places in the last two decades, no new extensive delimitation has been made. Parts of the zonal gradation, however, have been outlined in far more detail than before, in particular by Stone. After an extensive perusal of maps and photographs of the polar settlement fringes in North America and Scandinavia, he comes to the conclusion that it is possible to subdivide the area of discontinuous settlement into four "fringe zones" of varying isolation of their settlers, the amount increasing toward the anecumene (Fig. 8). Isolation is measured by the distance between the settlers and their accessibility by vehicles. Something similar to this concept of zonal gradation is offered by Ehlers, who, in his study of the settlement fringe in the Peace River area inserts, between the zone of scattered agricultural islands—his "young pioneer area" (*jung Pionierraum*)—and the compact farmlands, a "transition area" (*Übergangsraum*) where cultivated land still occupies a minority of the surface but occasionally comprises larger and more compact areas.[8]

Whether a detailed subdivision on the order of Stone's can also be recognized along the still more fragmented agricultural margins of the dry and mountain lands is another question, although the ability of Hambloch, in his studies of height limits in the Western United States, to differentiate between islands of agriculture generated by permanent settlements and those caused by temporary settlements would seem to

[7] E. Ehlers, *Das nördliche Peace River Country, Alberta, Kanada* (Tübingen: Univ. Tübingen, 1965) pp. 99, 127, 131, 134, 163, 165, 173, 179; J. W. Maxwell, "Notes on Land Use and Landscape Evaluation in a Fringe Area of the Canadian Shield," *Geographical Bulletin,* Vol. 8 (1966), p. 143, Fig. 14; W. R. Mead, "Cold Farm in Finland: Resettlement of Finland's Displaced Farmers," *Geographical Review,* Vol. 41 (1951), p. 535, Figs. 6, 7; M. Born, "Anbauformen an der agronomischen Trockengrenze Nordostafrikas," *Geographische Zeitschrift,* Vol. 55 (1967), 243–78; E. Ehlers, "Das boreale Waldland in Finnland und Kanada als Siedlungs- und Wirtschaftsraum," *ibid.*, Vol. 55 (1967), 280–322.

[8] I. Bowman, *The Pioneer Fringe* (New York: American Geographical Society, 1931), p. 50; K. J. Pelzer, "Limits of Settlement Possibilities," in *Limits of Land Settlement,* ed. I. Bowman (New York: Council on Foreign Relations, 1937), facing p. 380; F. Jaeger, *Die klimatischen Grenzen des Ackerbaus* (Zurich: Schweizerische naturforschende Gesellschaft, 1946), cited by Fels, *Der wirtschaftende Mensch,* 1st ed. (1954), p. 137. K. H. Stone recapitulates his basic conclusions in "Geographic Aspects of Planning for New Rural Settling in the Free World's Northern Lands," in *Problems and Trends in American Geography,* ed. S. B. Cohen (New York: Basic Books, 1967), 221–37. E. Ehlers, *Das nördliche Peace River Country,* p. 191.

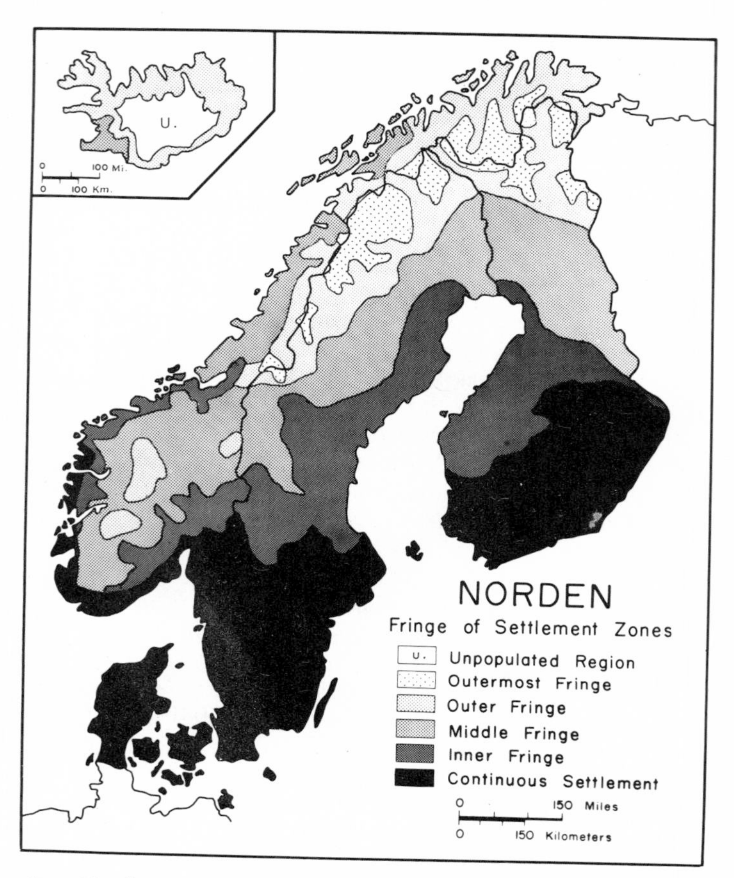

Figure 8. *Gradations in farming intensity on the agricultural frontier in Norden, as reflected in the isolation of farms. [Adapted from "Geographic Aspects of Planning for New Rural Settling in the Free World's Northern Lands," by Kirk H. Stone, p. 226 of* Problems and Trends in American Geography, *edited by Saul B. Cohen, Basic Books, Inc., Publishers, New York, 1967.]*

offer hope (Fig. 9). Hambloch, who includes an uninterrupted agricultural zone on the lower slopes of his "highland boundary fringe" (*Höhengrenzsaum*) takes a broader view than Stone of what constitutes the farming fringe, however.[9]

Additional disagreements over the delimitations of the agricultural fringe will probably develop if the divergence between settlement and agricultural margins continues to grow. Both margins have often been assumed to be practically synonymous, at least for general purposes, since the settlements lying beyond the agricultural foreposts and supported by nonagricultural activities are usually smaller and less populous than the

[9] H. Hambloch, "Höhengrenzen von Siedlungstypen in Gebirgsregionen der westlichen USA," *Geographische Zeitschrift*, Vol. 55 (1967), 1–41.

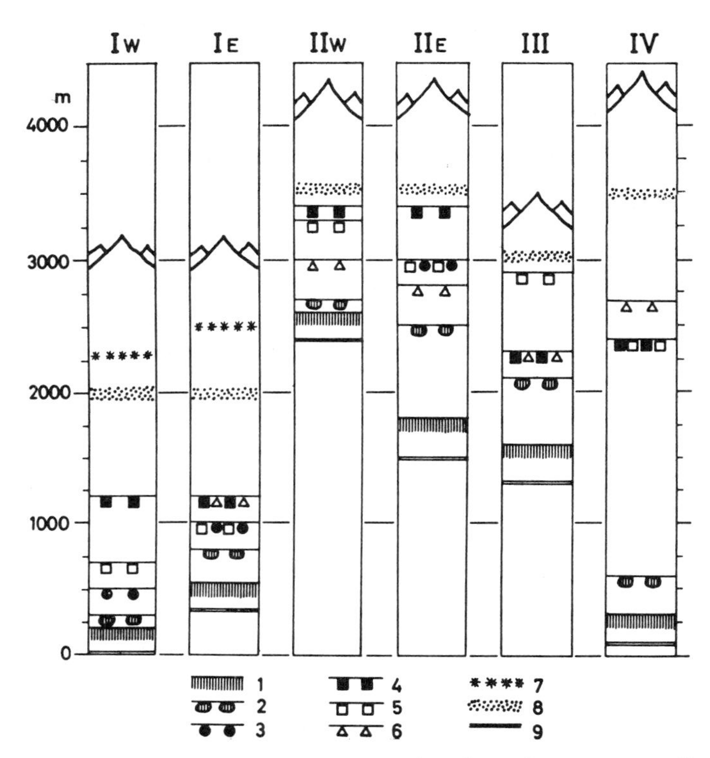

Figure 9. *Altitudinal boundaries of agricultural settlements in relation to those of other types in the Western United States.*
Iw, Ie: *parts of the western and eastern flanks of the northern Cascades;* IIw, IIe: *western and eastern flanks of the Colorado Front Range;* III: *the flanks of the Gallatin-Absaroka Range in Montana;* IV: *the western flanks of the southern Sierra Nevada.*
Boundaries of permanent settlements include those of: (1) *compact agricultural areas;* (2) *fragmented agricultural areas;* (3) *mining and hydroelectric installations; and* (4) *recreation and transportation settlements. The boundary of the seasonal settlements* (5) *applies to recreation and to transportation–maintenance groups. Boundaries of ephemeral settlements include those of:* (6) *fragmented agricultural areas;* (7) *all inhabited areas (snow boundary); and* (8) *forested areas (forest boundary). Lower limits of the mountain forelands are indicated by* (9). *[By permission of H. Hambloch and the* Geographische Zeitschrift, *55 (1967), 36.]*

farming outliers. But now these nonagricultural activities are becoming more important, and often precisely along the very farming margins that have become stationary or are even retreating. Logan has noted that environmental amenities and industrialization continue to cause urban tracts to proliferate in the deserts of southern California, Nevada, and Arizona. They have no agricultural base, and some are over a hundred miles from major urban centers. Hambloch also describes the burgeoning of permanent recreational settlements well above the agricultural islands in the Rockies.[10]

A similar gap is occurring along many parts of the polar agricultural margin in North America, but here it is more because of the establishment of mining towns in the wilderness and the retreat of agriculture from areas around older marketing centers. Many of these older towns continue to grow because of new industrial plants and increasing mineral discoveries, while the improvement of transportation to service these centers paradoxically weakens the ability of local farmers to compete with producers in more productive environments. Unfortunately, little of this development has yet been mapped, although the works of Krenzlin in intermountain British Columbia and of Clibbon and Gagnon in Quebec, along the northern edge of the St. Lawrence Valley, are excellent examples of what one hopes to see more of.[11]

Interregional Boundaries

Geographers have also been vitally interested in delimiting agricultural regions within the well developed farming areas. Here the ratio has been found to be the most efficient tool in determining where a boundary lies. It has been a major basis of the worldwide regional schemes of Engelbrecht, Baker, Hartshorne and Dicken, and Whittlesey, as well of the more detailed boundary studies of just one region, such as those by Hartshorne, Prunty, and Roepke of various American agricultural regions.

These, and other writers about individual regions evidence the same intense concern over the limits and character of the transition zones that occupies those who write about the agricultural fringe. Some have concentrated on outlining the zones of production intensity gradation that parallel the boundaries, like Hartshorne, who studied the Dairy Belt; Prunty, who considered the Cotton Belt; and Roepke, who wrote on the Corn Belt. Other writers have characterized transition zones more in terms of farming types. The outlining by McCarty of the "inner" and "outer" corn belts and "old" and "new" wheat belts in the American heartland may be especially mentioned, as well as the articles by Durand on those Dairy Belt subregions that owe their individuality to their transitional character. Durand has also probed the even more difficult problem of recognizing transitional zones between functional regions, by

[10] R. F. Logan, "Suburbia in the Sun: The Desert Boom in the Southwest," (Abstract), *Annals of the Association of American Geographers,* Vol. 46 (1956), 259; Hambloch, "Höhengrenzen von Siedlungstypen."

[11] A. Krenzlin, *Die Agrarlandschaft an der Nordgrenze der Besiedlung im intermontanen British Columbia* (Frankfurt am Main: Kramer, 1965), p. 30; P. B. Clibbon and J. Gagnon, "L'évolution récente de l'utilisation du sol sur la rive nord du Saint-Laurent entre Québec et Montréal," *Cahiers de géographie de Québec,* Vol. 10 (1956–66), No. 19, 55–72.

mapping the "competitive overlaps" of the major milksheds in the Northeastern United States.[12]

Not all geographers are convinced that agricultural boundaries and transition zones are so extensive. Some, like Gibson, have felt that if studies were more localized and detailed, the areas considered transitional would be reduced considerably. Others ask for less stress on both boundaries and transitional zones to begin with. Thus, Rühl maintains that these areas are of less importance to the world economy than the core areas, and Buchanan argues that delimiting studies ignore the economic conditions that brought about the regional specialization.[13]

[12] R. Hartshorne, "A New Map of the Dairy Areas of the United States," *Economic Geography,* Vol. 11 (1935), 347–55; M. Prunty, Jr., "Recent Quantitative Changes in the Cotton Regions of the Southeastern States," *ibid.,* Vol. 27 (1951), 189–208; H. G. Roepke, "Changes in Corn Production on the Northern Margin of the Corn Belt," *Agricultural History,* Vol. 33 (1959), 126–32; H. H. McCarty, *The Geographic Basis of American Economic Life* (New York: Harper, 1940), pp. 228–31, 294–308; L. Durand, Jr., "Cheese Region of Northwestern Illinois," *Economic Geography,* Vol. 22 (1946), 24–37; *idem,* "The Lower Peninsula of Michigan and the Western Michigan Dairy Region: A Segment of the American Dairy Region," *ibid.,* Vol. 27 (1951), 164–82; *idem,* "The Major Milksheds of the Northeastern Quarter of the United States," *Economic Geography,* Vol. 40 (1964), 9–33.

[13] L. E. Gibson, "Characteristics of a Regional Margin of the Corn and Dairy Belts," *Annals of the Association of American Geographers,* Vol. 38 (1948), 270; A. Rühl, *Das Standortsproblem in der Landwirtschafts-Geographie (Das Neuland Ostaustralien),* Veröffentlichungen des Instituts für Meereskunde, Neue Folge, B: Historischvolkswirtschaftliche Reihe, Heft 6 (Berlin: Mittler, 1929), p. 113; R. O. Buchanan, "Some Reflections on Agricultural Geography," *Geography,* Vol. 44 (1959), 12.

PART 4 *the concern for resources*

CHAPTER TEN

Resource destruction

The Destruction of the Soil

Nowhere is the misuse of resources so evident as in the agricultural landscape. The naturally extractive character of the plant-growth process, the long and often negligent use of the land by man, and the wide areas affected all have combined to make the problem a glaring one. Thus it has also become an object of interest to numerous workers in agricultural geography, often as part of larger studies but frequently also as a major topic in itself. Research in soil conservation problems is now so detailed and prolific in the United States that here it is considered a field.

It is probably only natural that soil destruction in this country should have concerned so many geographers, pedologists, soil conservationists, ecologists, and others. Soils have been eroded and exhausted more quickly and over a larger area in America than in any other similar-sized place in modern times. Moreover, much of the damage has seriously affected a large part of one of the very few agricultural regions in the world that are both fertile and extensive. The particularly devastating wind erosion of the Great Plains in the 1930's seems to have been the catalyst for a number of extensive writings, the most impressive being Bennett's massive volume, *Soil Conservation.* His judicious combination of survey, analysis, and prescription also became the policy of a federal agency, the Soil Conservation Service, created in 1935. This was the decade, too, when many of the principles developed by academic geographers in their surveys of derelict rural lands in the Great Lakes cut-over areas found national application in federal programs. None of this means that authors of other times and other countries have not contributed to the literature on land depredation, however. Book-length treatments of the world situation were begun at least a century ago, and

English, French, and German scholars have written impressive tomes as well.[1]

Much of the present research on soil erosion is now being done by agricultural, geological, and engineering specialists in governmental agencies. Particularly large numbers of such specialists in the United States Soil Conservation Service have been engaged in regionalizing "soil problem" areas on the basis of land use capability. Both governmental and academic researchers, however, have also inquired into a wide range of additional problems related to cultivation and its result on the soil. How rapidly American soil is eroding and what is the best method of measurement of the denudation rate has long interested hydraulic engineers like Leopold. In a related problem, the determination of the rate of fertility loss, Albrecht and other soil scientists appear to have been more successful in their discovery that the life-curve of a productive soil duplicates the death-rate curve, or loss of energy, of a radioactive chemical element from the moment forward of its creation by atomic explosion. Geographers, predictably, show up more in the discussions of regional and spatial aspects of soil erosion. Johnson displays a traditional geographical concern about the way the rectangular survey system has prevented better adaptation of cultivation to the terrain, in her description of the ways Minnesota farmers have juggled their land purchases to overcome this obstacle. One particular type of land form–field form discordance and the resultant erosion, that of the intersection of drumlins and field patterns at oblique angles, has especially attracted Chapman and Putnam in Ontario and Durand in Wisconsin. And in the Great Plains, the discrepancies between mean and actual precipitation values and their heightening of the danger of wind erosion has attracted Ahnert's attention.[2]

This unanimity of concern about soil erosion has by no means meant agreement about exactly how critical the problem is, or about how much of it has been caused by man. Opinions on these issues range

[1] H. H. Bennett, *Soil Conservation* (New York: McGraw-Hill, 1939). C. O. Sauer was another academician of influence; see, e.g., "The Problem of Land Classification," *Annals of the Association of American Geographers,* Vol. 11 (1921), 3–16. Perhaps the earliest book on soil destruction was G. P. Marsh, *Man and Nature: Or, Physical Geography as Modified by Human Action* (New York: Scribner, 1864); the latest, as of this writing, is E. Fels, *Der Wirtschaftende Mensch als Gestalter der Erde,* 2d ed. (Stuttgart: Franckh, 1967).

[2] A summary of much of the effort of the Soil Conservation Service is provided in M. E. Austin, *Land Resource Regions and Major Land Resource Areas of the United States,* Dept. of Agriculture Handbook No. 296 (Washington, D.C.: U.S. Govt. Printing Office, 1965).

L. B. Leopold, "Land Use and Sediment Yield," in *Man's Role in Changing the Face of the Earth,* ed. W. L. Thomas, Jr. (Chicago: Univ. of Chicago, 1956), pp. 639–47; W. A. Albrecht, "The 'Half-Lives' of Our Soils," *Natural Food and Farming,* Vol. 13 (1966), 6–10; H. B. Johnson, "Rational and Ecological Aspects of the Quarter Section: An Example From Minnesota," *Geographical Review,* Vol. 47 (1957), 339; L. J. Chapman and D. F. Putnam, *The Physiography of Southern Ontario,* 2d ed. (Toronto: Univ. of Toronto, 1966), pp. 77, 286; L. Durand, Jr., "Cheese Region of Southeastern Wisconsin," *Economic Geography,* Vol. 15 (1939), 283–92; F. Ahnert, "Neue Winderosionsgefahr im Weizenanbaugebiet der Great Plains," *Erdkunde,* Vol. 9 (1955), 217–20.

widely and cut across many lines of specialization. Few have echoed the dire warnings of Vogt and Osborn, and shared their despairing assessment of soil erosion, although Shantz has much the same view of the future condition of the soil in the drier lands. These writers believe that the trend toward soil destruction will be practically impossible to reverse, in view of the increasing pressure of population. Only radical and extensive population control, according to Vogt, can meet the problem posed by these "twin evils." Far more researchers, however, reject these views, and none more spiritedly than investigators of Marxian convictions. Soviet geographers, in particular, have long extolled the ability of man to "transform nature," and Davitaya has answered Schantz directly, claiming that proper exploitation can continually enrich the soils of arid lands. Even the radiation balance, he maintains, "a factor apparently so conservative and so little dependent upon man, . . . can, with irrigation, be increased in the deserts by several dozen per cent from its original magnitude." As for population pressure, Marxists see it only as a consequence of poor exploitation of resources and inadequate distribution.[3]

Between these extremes of gloom and aggressive optimism can be found the largest group of opinion, that the problem of soil resources can be viewed with guarded optimism. Although the emphasis is on the vast potentialities that man has for the solving of the problem, the argument is commonly couched in illustrations of how he has destroyed the soils, and thus also of what might happen if he continues to ignore the consequences. Moreover, he must work with nature and not against it, as Whitaker asserts in his analysis of the Soviet view, for "programs of land reclamation, may, . . . actually do more damage than good." Though this contradistinction of potentialities for disaster and success is the dominant theme, variations in stress are not absent. More topically oriented specialists, like the soil scientist Bennett, seem more assured: "We have the knowledge; we need only the decision." Geographers, perhaps more sensitive to man's whimsy in his treatment of the landscape, have tended to view things more reservedly, although even here strong exceptions can be noted. Certainly, few researchers in other fields can surpass in pragmatic optimism and energetic approach a work like that of Smith on the possibilities of tree crops on lands highly prone to erosion.[4]

Still another view of the problem of soil erosion involves a curious

[3] W. Vogt, *Road to Survival* (New York: William Sloane Associates, 1948); E. P. Osborn, *Our Plundered Planet* (Boston: Little, Brown, 1948); H. L. Shantz, "History and Problems of Arid Lands Development," in *The Future of Arid Lands*, ed. G. F. White (Washington, D.C.: American Association for the Advancement of Science, 1956), pp. 3–25; F. F. Davitaya, "Transformation of Nature in the Steppes and Deserts," in *Soviet Geography: Accomplishments and Tasks*, ed. I. P. Gerasimov, trans. L. Ecker, ed. Eng. edition C. D. Harris (New York: American Geographical Society, 1962), p. 281.

[4] J. R. Whitaker, "World View of Destruction and Conservation of Natural Resources," *Annals of the Association of American Geographers*, Vol. 30 (1940), 146; H. H. Bennett, "Adjustment of Agriculture to Its Environment," *ibid.*, Vol. 33 (1943), 163; J. R. Smith, *Tree Crops: A Permanent Agriculture* (New York: Devin-Adair, 1950).

blend of the most optimistic and pessimistic appraisals. This is the belief that destructive exploitation will force man eventually to take protective measures, and that overexploitation is necessary because only its disastrous effects will make him realize that new measures must be taken. This kind of historical determinism was early expressed in Friederich's three-stage sequence: (1) ruthless exploitation; (2) lands in want; and (3) lands with soils conserved or improved. More recently Bates has formulated a sequence of exploitation—conservation—reconstruction, and Andreae has correlated the "development stages" (*Entwicklungsstufen*) of a national economy with changes in soil treatment by livestock farmers. Other writers, like Whitaker, Glendinning and Torbert, and Sauer, have rejected this kind of historical rationalization. They have been quick to assert that a vicious cycle of soil destruction can also take place, "in which depletion leads to further depletion, and in which want leads to further exploitation and eventually, perhaps, to complete soil removal." Most recently, Gregor has professed to see a similarly destructive cycle operating among irrigation farmers, with new means of extracting water encouraging faster-falling water tables, which, in turn, can only be remedied by still more technological advancements.[5]

Not all scholars have accepted the thesis that man is largely responsible for most of the devastating soil erosion. There is clear evidence of rapid erosion during periods of drought in recent geological time but before the known advent of man, and some conclude that the same thing is happening today. Clark claims that climatic fluctuations and other natural causes, and not the commonly cited overgrazing of man's herds and flocks, have been the principal reasons for the deterioration of productive forage and subsequent erosion of the Great Plains. Küchler also emphasizes that the effects of a dry cycle, and not necessarily the ravages of overgrazing, leads to degradation of extensive grass covers. Indirect support for the primacy of natural factors has also come from writers like Huntington, who sees a close correlation between the abandonment of land and climatic cycles in North Africa and Southwest Asia.[6]

The resistance to this milder view of man's role in erosion has been strong, particularly when it has been a question of a drought which ex-

[5] E. Friedrich, "Wesen und geographische Verbreitung der 'Raubwirtschaft,'" *Petermanns Mitteilungen,* Vol. 50 (1904), 68–79, 92–95; M. Bates, *Where Winter Never Comes,* 2d ed. (New York: Scribner's, 1963), p. 59; B. Andreae, *Weidewirtschaft im südlichen Afrika* (Wiesbaden: Steiner, 1966), pp. 31–33; Whitaker, *op. cit.,* pp. 155 f.; R. M. Glendinning and E. N. Torbet, "Agricultural Problems in Grainger County, Tennessee," *Economic Geography,* Vol. 14 (1938), 166; C. O. Sauer, "Destructive Exploitation in Modern Colonial Expansion," in *Comptes rendus du congrès international du géographie Amsterdam,* Vol. 2, Sec. III-c: *Géographie coloniale* (Leiden: Brill, 1938), pp. 496–99; H. F. Gregor, "Water and the California Paradox," in *The California Revolution,* ed. C. McWilliams (New York: Grossman, 1968), p. 164.

[6] A. H. Clark, "The Impact of Exotic Invasion on the Remaining New World Mid-latitude Grasslands," in *Man's Role in Changing the Face of the Earth,* pp. 737–62; A. W. Küchler, *Manual to Accompany the Map, Potential Natural Vegetation of the Conterminous United States* (New York: American Geographical Society, 1964), pp. 22–23; E. Huntington, *Mainsprings of Civilization* (New York: Wiley, 1945; Mentor Books, 1959), pp. 536–69.

tends over decades. Thus Butzer and Bobek belittle the thesis of long-term desiccation advocated by Huntington and others, and instead believe that destructive wars and the resultant neglect of the soil have played the decisive role. The archeologist Braidwood also sees no evidence to contradict his feeling that little environmental change capable of changing the agricultural scene has occurred in Southwestern Asia during historic times. Man also appears to be the chief culprit to Sauer, who raises the question of why the western Sahara is much more desolate than its climatic homologs, Lower California and Sonora. A longer period of stocking in the Old World seems to him to be the principal reason.[7]

Recent research has added new depth to the debate over the relative importance of the roles of man and nature in erosion and has suggested a more complicated answer. Schoenwetter and Eddy have contradicted those who have claimed that drought is the principal destroyer of soil in the American Southwest, shifting the blame to periods of intense summer rainfall. This assertion, insofar as prehistoric gullying is concerned, is based primarily on the interpretation of evidence from pollen, rather than on the evidence from tree rings on which the upholders of the older theory rely. Tuan has offered support for this newer theory with his several correlations between modern gullying and intense storms and flood runoff. A perusal of the history of livestock population in the same area, however, has persuaded Denevan that emphasis on a particular agent of erosion, be it drought, heavy rains, or man's livestock, is misleading. His comparisons of livestock "grazing pressure" with changes in the intensity of rainfall and arroyo cutting provide a defense for both climatic theories, as well as suggesting the plausibility of overgrazing.[8] Presumably, then, investigation should be concentrated less on finding a single, dominant agent of erosion, and more on determining the proper perspective of the major or minor roles each agent plays.

Urban Encroachment

Another well-developed research concern of those investigating agricultural resources has been the loss of farmland to expanding cities. Geographers, economists, and soil scientists have all contributed to the

[7] K. W. Butzer, "Climatic Change in Arid Regions since the Pliocene," in *A History of Land Use in Arid Regions,* ed. L. D. Stamp (Paris: UNESCO, 1961), pp. 44 f.; H. Bobek, *Vegetationsverwüstung und Bodenerschöpfung in Persien und ihr Zusammenhang mit dem Niedergang älterer Zivilisationen* (Vienna: Geographisches Institut, Univ. Wien, 1958); R. Braidwood, "Jericho and Its Setting in Near Eastern History," *Antiquity,* Vol. 31 (1957), 73–81; C. O. Sauer, *Agricultural Origins and Dispersals* (New York: American Geographical Society, 1952), pp. 101–103; *idem,* "The Agency of Man on Earth," in *Man's Role in Changing the Face of the Earth,* pp. 60 f.

[8] J. Schoenwetter and F. W. Eddy, *Alluvial and Palynological Reconstruction of Environments* (Santa Fe: Museum of New Mexico Press, 1964); Y. F. Tuan, "New Mexican Gullies: A Critical Review and Some Recent Observations," *Annals of the Association of American Geographers,* Vol. 56 (1966), 573–97; W. M. Denevan, "Livestock Numbers in Nineteenth-Century New Mexico, and the Problem of Gullying in the Southwest," *ibid.,* Vol. 57 (1967), 691–703.

literature. The record of observations is relatively short, but this is because the rapid horizontal growth of cities has been a major worldwide phenomenon only since World War II. Like the work on soil erosion, however, the largest part of the research on urban encroachment on agricultural land has been done in the United States, again because it has been most dramatic here.

Although the primary stimulant of the swelling amount of writing on the problem is the simple fact of diminishing farm land, most of the attention has focused on three particular spatial aspects of this diminishment: selective encroachment; fragmentation of agricultural land; and "urban shadow," the effects of the advancing city on adjacent agricultural land. Relatively heavier losses of the best soils to urbanization have been reported by many researchers. An especially detailed study by Krueger has revealed that the Niagara Fruit Belt of Ontario, the major fruit area of Canada, and now in the path of rapidly advancing cities, is also that part of the province where the ratio of "tender fruit soils" to "tender fruit climate" is precisely the most favorable. In California, the author has noted a particularly strong tendency toward urbanization of the better soils, undoubtedly in good part because of the more limited areas of gentle terrain available to both cities and agriculture, and because so many of California's urban centers originated as service centers for farming communities. Specific correlations between urban thrusts and particularly fertile one-crop areas have also been described—for instance, Gregor on orange groves in southern California, and Psuty and Salter on mango groves in southern Florida.[9]

The fragmentation of agricultural land, rather than just encroachment on its edges, has been studied intensely because it creates a greater disruption of farming activities and accelerates the rate of land transfer. Fragmentation exposes many more farmers to a whole range of problems than does a more compact urban advance, foremost of which is an increase in tax assessments designed to pay for the growing urban population nearby and to reflect the heightened possibility that the remaining agricultural lands can be converted to urban uses. Fragmentation has also been called "atomization" because it leads to an ever-widening and scattered pattern of land use influences prejudicial to farming. Real estate developers are constantly in search of cheap land, but in finding and establishing their developments they create still more high-price land, which in turn encourages still more forays into other cheap areas. The author has found such a process particularly devastating in California. Durand hints at a possible double irony in this process when he relates how operators of larger farms commonly join in the competition for

[9] R. R. Krueger, "Changing Land-use Patterns in the Niagara Fruit Belt," *Transactions of the Royal Canadian Institute*, Vol. 32 (1959), 39–140; H. F. Gregor, "Urban Pressures on California Land," *Land Economics*, Vol. 33 (1957), 311–25; *idem*, "Urbanization of Southern California Agriculture," *Tijdschrift voor Economische en Sociale Geografie*, Vol. 54 (1963), 275; N. P. Psuty and P. S. Salter, "Land Use Competition on a Geomorphic Surface: The Mango in South Florida," *Annals of the Association of American Geographers*, Vol. 59 (1969), 264–79.

cheap land in order to obtain a buffer between themselves and the subdivisions; yet the consequent scattered distribution of the large farmers themselves may well lead one to conclude that the over-all areal result of this bidding will often be to stimulate greater dispersal of the urban fragments and their detrimental effects on nearby agriculture.[10]

For many scholars, urban encroachment has been demanding of attention primarily because of its striking effects on the immediately adjacent agricultural landscape. The decline in farming activity brought about by high urban taxes, the anticipation of urbanization, and other urban-caused factors has become so widespread around North American cities that it has convinced Sinclair it is now possible to see a rough ring-like zonation of agricultural uses around a city, their intensity decreasing as the city is approached (see Chap. 4). Others, however, have observed the common increase in the intensification of farming in the remaining crop patches on the urban margins, a reaction which Gaffney claims has led to a contradictory agricultural resource problem—excess capacity.[11]

The sprawl of cities over agricultural land has evoked a wealth of proposals for the retention of the choicest acreages for farming. Most believe that restrictive zoning is the best solution, Solberg being its most ardent advocate. The vexing problem facing planners, of how to decide whether a piece of land should remain agricultural or be allowed to become urban, has been attacked several ways. The most promising step toward a solution so far seems to be the suggestion by Wibberley and Ward, in Britain, of benefit-cost analysis, in which alternative areas available for urban development would be evaluated on the basis of which would be least costly to the community. The significant costs would include the net agricultural values foregone and the costs of essential improvements. Others feel that agricultural zoning is no longer the answer, and press for decisive national efforts. The most drastic proposal yet made in the United States has probably been Griffin and Chatham's. They suggest that the federal government acquire title to agricultural land and reserve it for that use. It would then be sold or leased to operators, and the restrictions on its use stipulated in the deeds.[12]

[10] H. F. Gregor, "Spatial Disharmonies in California Population Growth," *Geographical Review,* Vol. 53 (1963), 119; L. Durand, Jr., "The Retreat of Agriculture in Milwaukee County, Wisconsin," *Wisconsin Academy of Sciences, Arts and Letters,* Vol. 51 (1962), 200.

[11] R. Sinclair, "Von Thünen and Urban Sprawl," *Annals of the Association of American Geographers,* Vol. 57 (1957), 72–87; M. Gaffney, "Containment Policies for Urban Sprawl," in *Approaches to the Study of Urbanization,* ed. R. L. Stauber (Lawrence: Univ. of Kansas, 1964), pp. 2 f.

[12] E. D. Solberg, *Talks on Rural Zoning,* Dept. of Agriculture, Agricultural Research Service (Washington, D.C.: U.S. Govt. Printing Office, 1960); G. P. Wibberley, *Agriculture and Urban Growth: A Study of the Competition for Rural Land* (London: Michael Joseph, 1959), pp. 73–101; J. T. Ward, "The Siting of Urban Development on Agricultural Land," *Journal of Agricultural Economics,* Vol. 12 (1957), 451–66; P. F. Griffin and R. L. Chatham, "Urban Impact on Agriculture in Santa Clara County, California," *Annals of the Association of American Geographers,* Vol. 48 (1958), 207.

The alarm over the loss of agricultural land, even of the best, is not uniformly shared. Harris, an urban geographer, has argued that the role of cities as centers of culture and economic change is of much more weight than their role as competitors of agriculture for space. Considering all the benefits provided by the cities, including their inestimably greater capacity for supporting people through jobs and their production of farm machinery and other technological means for improving farm production, Harris sees in the arbitrary reservation of all good farmland only a lowered standard of living. Nor does he foresee any major loss of total world agricultural land to cities, either now or in the future. Hart agrees, and maintains that urban encroachment, even of the largest cities in the Eastern United States, has been only a minor inducement to land abandonment in comparison with other stimuli. Economists like Gillies, Mittelbach, and Lessinger also see no problem of misallocation of resources so long as the forces of supply and demand are allowed to operate freely. Thus presumably a time will come when agricultural land will become so scarce that prices offered for it by those interested in farming will exceed those offered by urban interests, and the land will stay in agricultural use. A less common objection to those who fear irreparable damage to agricultural productivity by urban sprawl is raised by Gaffney, who asserts that the problem is one of overintensification, and thus of surplus production, in reaction to higher land prices.[13]

Few disagreements of this kind have accompanied other resource problems associated with the growth and spread of cities over agricultural land, principally because they have been less obvious and their damages are not "balanced" by an apparent gain elsewhere. Soil compaction and the sinking of land surfaces, with a consequent reduction of soil-water capacity and an increase in the danger of floods comprises one of these problems. Damage to plants from air pollution is another. The most critical of these additional problems, however, is the growing competition between city and rural areas for the limited water supplies in arid regions, particularly the Southwestern United States. Clawson, in fact, foresees that the major cities in these regions will take all the available water and all the most suitable land areas in their environs within a generation. None of this can happen too soon, in the view of writers like Wilson, who points to the large amounts of water consumed by an agriculture that supports but a fraction of the labor force in the oases. Countering this view is the more cautious one of Gregor, who raises the question of whether the sacrifice of the important specialty-crop areas of the West is worth the further promotion of industries that are still only a small part

[13] C. D. Harris, "The Pressure of Residential-Industrial Land Use," in *Man's Role in Changing the Face of the Earth*, pp. 881–95; J. F. Hart, "Loss and Abandonment of Cleared Farm Land in the Eastern United States," *Annals of the Association of American Geographers*, Vol. 58 (1968), 426–29.

J. Gillies and F. Mittelbach, "Urban Pressures on California Land: A Comment," *Land Economics*, Vol. 34 (1958), 80–83. A reply to this article is made by H. F. Gregor in the same issue, " 'Urban Pressures on California Land': A Rejoinder," pp. 83–87.

J. Lessinger, "Exclusive Agricultural Zoning: An Appraisal, I. Agricultural Shortages," *ibid.*, Vol. 34 (1958), 149–60, 255–62; Gaffney, *op. cit.*

of the national industrial plant and are heavily oriented toward military needs.[14]

The Human Resource

Concern for the human resource underlies all studies of threats and damage to agricultural resources, but in some of these works that worry expresses itself more directly. Vogt and Sauer, for example, have stressed the close link between soil erosion and population capacity. Vogt believes that the very ability of man to survive is threatened by his uncontrolled pressure on the land; to Sauer, the significance of land as a resource of man is justification for a "developmental classification of regions according to economic conditions," the most critical regional type being "areas of chronic crisis," where destructive exploitation has left population in excess of the sources of subsistence that remain. The effect of soil depletion on diet is another facet of the man-land relationship. Albrecht warns of an increasing discrepancy between the quality and the bulk of food in the United States, the price of mining the fertility of the soil. Kellogg considers mineral deficiencies serious enough to destroy cultures and races.[15]

Conflicts between human groups have also been closely associated with soil. That wars lead to devastation of the soil has been a favorite argument of those who see most such damage as a result of man. A related argument is that the soil has been destroyed by colonial exploitation. Few have matched the vehemence of Marsh when he blamed imperial Rome for much of the "sterility and physical decrepitude" that is still evident in much of the land it once ruled, although Sauer and Harroy are certainly in agreement. Both have condemned European expansion for extensive impoverishment of the lands colonized, Sauer largely on the basis of his observations in Latin America, Harroy in Africa.[16]

The condition of rural man as affected by the swelling migration of farmers into the cities has also been viewed with alarm by a good num-

[14] The mapping by P. A. Leighton of the present and potential areas of plant damage by air pollution in California has been a major step in research on that problem; see "Geographical Aspects of Air Pollution," *Geographical Review*, Vol. 56 (1966), 166.

M. Clawson, "Changing Patterns of Land Use in the West—I," in *Resources Development: Frontiers for Research*, ed. F. S. Pollak (Boulder: Univ. of Colorado, 1960), pp. 217–28; A. W. Wilson, "Urbanization of the Arid Lands," *Professional Geographer*, Vol. 12 (1960), 7; H. F. Gregor, "Competition for Rural Land: A Rationale for Planning," in *Spatial Organization of Land Uses: The Willamette Valley*, ed. J. G. Jensen (Corvallis: Oregon State Univ., 1964), p. 37.

[15] Vogt, *op. cit.;* C. O. Sauer, "Land Resource and Land Use in Relation to Public Policy," *Report of the Science Advisory Board: July 31, 1933, to September 1, 1934*, Appendix 9 (Washington, D.C.: U.S. Govt. Printing Office, 1934), pp. 249 f.; W. A. Albrecht, "Soil Fertility and Biotic Geography," *Geographical Review*, Vol. 47 (1957), 86–105; C. E. Kellogg, "Soil and the People," *Annals of the Association of American Geographers*, Vol. 27 (1937), 142–48.

[16] Marsh, *op. cit.*, pp. 10–13; Sauer, "Destructive Exploitation," pp. 494–99; *idem*, "The Agency of Man," pp. 62–64; J. P. Harroy, *Afrique, terre qui meurt: La dégradation des sols africains sous l'influence de la colonisation* (Brussels: Marcel Hayez, 1944).

ber of writers. Many, like Huntington and Baker, have concluded that the movement is draining the rural areas of its best people and, in placing more of the population in an environment that encourages fewer children, may eventually threaten entire nations or even civilization itself with collapse. Much of this opinion, however, was reached before World War II, after which birth rates surged and agricultural surpluses began to form despite even sharper inroads on the rural population. No better reflection of these changes is provided than by the approaches to American farm migration by Baker in 1933 and Higbee in 1963. Baker recommended the relocation of industrial establishments in villages to stem the flow of farmers; Higbee advocated an even greater reduction of the rural population, calling for the shifting of government funds for crop support to the cities to take care of the now overwhelming urban majority. Many sociologists have also seriously questioned the thesis that rural-urban migration is selective. Some of the dissenters, like Smith, say it is possible that the city may attract large numbers of both the best and worst elements of the country, with the net result that the migration is neither advantageous to the city nor disadvantageous to the country.[17]

This more optimistic view of rural out-migration is still a tempered one, however. The turning down again of birth rates in the Western nations in the sixties has reinforced the feelings of those who still see population disaster for the West on the horizon. The increasing shift to urban environments is also causing writers to raise ever more questions about the psychological effects on migrants. In Europe, where the move means a definite cultural as well as occupational change, opinion, exemplified by Fels and Otremba, is generally gloomy, although a few, like Faucher and George, are cautiously optimistic about the ability of the remaining peasants to adjust to a more industrialized agriculture. Alarm is now also being expressed over the out-migrations of rural folk in the much more populous underdeveloped segment of the world. Most cities lack the industry to support all the newcomers, and a large urban proletariat appears in the making. Nor do all students of agriculture in underdeveloped areas feel that once the cities are eventually sufficiently industrialized, the machinery, tools, and fertilizers they will manufacture will stimulate agricultural production as it has in the Western countries. Boserup believes that there is little incentive, in primitive rural economies, to industrialize agriculture so long as food is cheap and industrial goods are dear. To increase agricultural prices, she would restrict migration from the countryside, an opinion based additionally on her conviction that past agricultural development in pre-industrial societies has

[17] Huntington, *op. cit.*, pp. 76–107; *idem*, "The Conservation of Man," in *Conservation of Natural Resources*, ed. G. H. Smith (New York: Wiley, 1950), pp. 466–83; O. E. Baker, "Rural-Urban Migration and the National Welfare," *Annals of the Association of American Geographers*, Vol. 23 (1933), 59–126; E. Higbee, *Farms and Farmers in an Urban Age* (New York: Twentieth Century Fund, 1963), pp. 142 f.; T. L. Smith, *The Sociology of Rural Life*, 3d ed. (New York: Harper, 1953), pp. 174 f.

been encouraged not so much by the introduction of improved techniques from the outside as by the pressure of population within.[18]

Less is known about the attitudes of rural man toward the various resource problems he must face, and much of what has been discovered is of fairly recent vintage. One major exception is the long study of changing perceptions of the drought hazard in the Great Plains, but even here most contributions have been byproducts of studies of other aspects of the region. Articles like Kollmorgen's, which traces the origin of the belief that increasing rainfall followed settlement, and Malin's, which delineates the phases that ideas about the climatic problem have gone through, are infrequent. The first large-scale systematic analysis of farmers' attitude toward drought was not made until 1966, when Saarinen obtained the responses of farmers to questionnaires and apperception tests and compared them with drought records. Six sample areas were selected within the winter wheat region of the central Great Plains, according to various specified degrees of aridity. Saarinen's principal conclusion was that farmers in the driest areas, as well as those who are older and have more experience in the region, are most perceptive of the drought hazard. Similarly, grain farmers are more perceptive of the danger than stockmen. Less expected, perhaps, are his findings that almost all farm operators underestimated the frequency of drought and that the very oldest farmers were as a group less perceptive than their experience would indicate.[19]

The differences among farmers' evaluations of the worth of a resource can also be numerous. How vital a thorough knowledge of these differences can be to those in charge of conservation planning has been explored in detail by Burton, in his study of the flood problem as associated with agricultural occupance of flood plains in the United States. He makes the crucial point that public planners facing the problem of minimizing flood danger all too commonly estimate the value of the land as it is, rather than as it is perceived by the farmer. Depending on how large the flood danger bulks in his mind and on the alternatives available to him, the farmer may react to the problem in a variety of ways. Measures to restrict cultivation on lands of high flood frequency, for example, may be stoutly resisted by farmers whose only arable land is on the flats. Nor can provision of flood protection necessarily be expected to improve significantly the farmer's view of the flood plain. Acreage controls in price-support programs, shortage of labor, lack of capital, and

[18] Fels, *op. cit.*, p. 216; E. Otremba, *Allgemeine Agrar- und Industriegeographie*, 2d ed. (Stuttgart: Franckh, 1960), pp. 197–99; D. Faucher, *Le paysan et la machine* (Paris: Editions de Minuit, 1954); P. George, *Précis de géographie rurale* (Paris: Presses univ. de France, 1963); E. Boserup, *The Conditions of Agricultural Growth* (Chicago: Aldine, 1965).

[19] W. M. Kollmorgen, "Rainmakers on the Plains," *Scientific Monthly*, Vol. 60 (1935), 146–52; J. C. Malin, *The Grassland of North America: Prolegomena to Its History with Addenda* (Lawrence, Kansas: Privately published, 1961); T. F. Saarinen, *Perception of the Drought Hazard on the Great Plains* (Chicago: Univ. of Chicago, Dept. of Geography, 1966).

many other restraints can contribute to his hesitation to expand the crop area.[20]

Studies of the relation of attitudes to the problems of soil erosion and urban encroachment are surprisingly few, considering the abundance of material on the subjects themselves. What has been done, however, provides a variety of leads for further research. One excellent example is the investigation by Blaut *et al.* of why soil conservation methods were not being accepted by farmers in the severely eroded Blue Mountains of Jamaica. Among the reasons found was that the farmers failed completely to perceive, or perceived imperfectly, the causes and effects of soil erosion. Investigating why urban encroachment proceeds more rapidly in some parts of the rural-urban fringe than in others, Birch found an areal variation in the farmers' appreciation of the opportunity costs created by the encroachment, these costs being the additional income that the farmer forgoes by not selling his land and equipment and investing the proceeds, and by not seeking other employment. But greater resistance to urban pressures may not always mean less cognizance of opportunity costs, as the author found in his search for an explanation of why many southern California farmers continue to hold onto their farms in the face of ever higher taxes and attractive financial offers. Pride in production and modern techniques, particularly when reinforced by a desire for farming as a way of life, as it is among certain ethnic groups, are the basic causes, in his estimation.[21]

[20] I. Burton, *Types of Agricultural Occupance of Flood Plains in the United States* (Chicago: Univ. of Chicago, Dept. of Geography, 1962).

[21] J. Blaut *et al.*, "A Study of Cultural Determinants of Soil Erosion and Conservation in the Blue Mountains of Jamaica," *Social and Economic Studies*, Vol. 8 (1959), 402–20; J. W. Birch, "On the Stability of Farming Systems, with Particular Reference to the Connecticut Valley," in *Essays in Geography for Austin Miller*, ed. J. B. Whittow and P. D. Wood (Reading, Eng.: Univ. of Reading, 1965), pp. 225–46; Gregor, "Urbanization of Southern California Agriculture," p. 277.

CHAPTER ELEVEN

Population and food supply

The Capacity of the Earth and Its Regions

This last chapter could well be the first in our survey of themes in agricultural geography, for some of the first noteworthy contributions to the literature were made by seventeenth and eighteenth-century political and agricultural economists concerned over the areal variations in the quality of land and their effect on the economic life of peoples. It was Malthus in particular who, in 1798, posed the problem in its most critical terms, namely as one of balance between the growing population and the capacity of the earth to produce food.

Economists, geographers, biologists, sociologists, and a host of other scientific, as well as popular, writers have written countless books, articles, and pamphlets about this problem. Despite this abundance of work and many differences in opinion, we can easily point out several major emphases and cite some of the more illustrative writings. The earliest aspect of the problem to be studied and the one which has received attention the longest, is the population capacities of the earth and its regions. Vague estimates of world population limits were already being made as early as the mid-eighteenth century, but it was not until 1890 that an attempt was made, by Ravenstein, an English geographer, to propose a figure based on a critical evaluation of the land. By estimating the relative areas in arable, steppe, and desert lands and then applying to them the agricultural results obtained in Britain, Ravenstein came to the conclusion that no more than about 5,995 millions could live on the planet and that this figure would be reached by the year 2072. Another critical element in the calculation of world population capacities, variation in living standards, was added in 1912 by Ballod. He translated the dietary demands of Americans, Germans, and Japanese into acreage requirements, applied these figures to his estimate of arable land acreage, and arrived at the respective maximum populations of 2,333; 5,600; and 22,400 millions.[1]

Since then, other important variables have been introduced into population calculations, although researchers are still not agreed about the relative importance of each agent. Penck stressed the role of climate, applying the highest population densities in each climatic area to the whole of the area. His resulting "highest conceivable" population for the earth was 15,900 millions. More than twice that amount, 33,000 millions, was proposed by Hollstein, who also emphasized the quality of climate but only insofar as arable land was available for its capitalization and minimum dietary requirements could be met. Full application of the discoveries in chemistry was the key, according to Rochow, who believed that it would then be possible to support 15,000 millions, one billion in the United States alone. Fischer felt that the political unit was the principal factor, and in his computations applied the maximum productivity figure for the country to all of its territory. His result was a qualified 6,200 millions, based solely on the ability of each country to support its population through its domestic resources and at the then-current level of its technology. Changing technological levels were the crux of Fawcett's estimates, however. He believed that if these levels were not raised, no more than 6,500 millions could live on the earth if French living standards became universal, and no more than 10,000 millions could survive if Indian living standards became general. If, on the other hand, expansion of agricultural land took place, especially in the tropics, and French living standards became general in the middle latitudes and Japanese standards common in the low latitudes, then Fawcett foresaw a maximum population of 9,600 millions. Finally, according to Fawcett, if all the potentialities for increases in acre-yields could be fully exploited, then the world's population could be easily doubled or tripled.[2]

Particularly good examples of population-capacity studies can also be found among the works of regional investigators. By computing the total deviations of areas from what he considered to be the values of ideal location, coal, temperature, and moisture, Taylor produced a figure of 1,347 millions of potential population in the chief area of white settlement. Baker postulated, in 1923, five different population ranges for the United States, depending on the extent of reclamation, yield intensifica-

[1] E. G. Ravenstein, "Lands of the Globe Still Available for European Settlement," *Proceedings, Royal Geographical Society,* Vol. 13 (1891), 27–35; K. Ballod, "Wieviel Menschen kann die Erde ernähren?" *Schmollers Jahrbuch für Gesetzgebung, Verwaltung und Volkswirtschaft* N. F., Vol. 36 (1912), 595–616.

[2] A. Penck, "Das Hauptproblem der physischen Anthropogeographie," *Zeitschrift für Geopolitik,* Vol. 2 (1925), pp. 330–48; W. Hollstein, *Eine Bonitierung der Erde auf landwirtschaftlicher und bodenkundlicher Grundlage,* Petermanns Mitteilungen, Ergänzungsheft 234 (Gotha: Haack, 1937). Rochow is cited in K. Scharlau, *Bevölkerungswachstum und Nahrungsspielraum* (Bremen-Horn: Walter Dorn Verlag, 1953), p. 131. A. Fischer, "Zur Frage der Tragfähigkeit des Lebensraumes," *Zeitschrift für Geopolitik,* Vol. 2 (1925), 762–79, 842–58; C. B. Fawcett, "On the Distribution of Population over the Land," *Social Review,* Vol. 17 (1925), 85–104; *idem,* "The Numbers and Distribution of Mankind," *Scientific Monthly,* Vol. 64 (1947), 389–96.

tion, export of farm products, and shift in dietary emphases. The lowest range was 105 to 125 millions, the highest, 400 to 500 millions. Some researchers have sought to establish a norm so as to be able to delineate areas of "over" or "underpopulation." Glendinning and Torbert, in eastern Tennessee, used a particular farm size and its productive capabilities to determine areas of stress among the rural population. Sedlmeyer and Müller-Wille, in their classification of "surplus" and "deficient" food production areas in Europe, used the average individual share of the continental harvest and the annual food requirements of the individual, expressed in the caloric production of the principal food crops, as indicators.[3]

Although the proliferation of these studies demonstrates strikingly well how provocative research can initiate a chain of reaction to a subject, the wide variation in population estimates and their lack of direct comparability (because of the different methods used) have destroyed their usefulness, in the eyes of many. Moreover, some of the estimates have already been exceeded. Countries like Japan and West Germany are enjoying the greatest prosperity in their history—with also the most crowded conditions. More recently, therefore, the answer to population capacities has been sought in optimum or relative, rather than maximum or absolute, terms. Further, the optimums are now viewed as dependent on a complex of economic, social, and psychological conditions, rather than on merely the productive capabilities of agricultural land.

The support for the newer view rests on the now-established fact that population does not grow steadily, as was assumed in most population estimates, but rises and falls. The fluctuations apparently represent the alternating effects of favorable and unfavorable historical events on the reproduction rate, so that optimum population levels can be obtained many times, although with never necessarily the same population numbers. Wagemann, a statistician, maintains that these changes have been regular enough to warrant the formulation of a "demodynamic alternation law" (*demodynamischen Alternationsgesetz*), in which each optimum correlates with a specific and progressively larger population density. Demographers Notestein and Thompson also advocate a cyclical pattern but, like most of their colleagues, refuse to make assertions as detailed as Wagemann's because of the generally scanty population records. Considerable data, however, exist for many Northwest European countries, and it is this material that forms the basis of their theory. Both were impressed by the impact of the industrial revolution on birth and death rates, the first effect being a rapid reduction of the death rate

[3] G. Taylor, "The Distribution of Future White Settlement: A World Survey Based on Physiographic Data," *Geographical Review*, Vol. 12 (1922), 375–402; O. E. Baker, "Land Utilization in the United States: Geographical Aspects of the Problem," *ibid.*, Vol. 13 (1923), 9–15; R. M. Glendinning and E. N. Torbert, "Agricultural Problems in Grainger County, Tennessee," *Economic Geography*, Vol. 14 (1938), 162–66; K. A. Sedlmeyer, "Die Tragfähigkeit des europäischen Lebensraumes und seine Ernährungsgrundlage: Auf Grund der Kulturpflanzenerträge," *Geographische Anzeiger*, Vol. 41 (1940), 233–45; W. Müller-Wille, "Europa-Seine Bevölkerung, Energieleistung und Ländergruppen," *Verhandlungen des deutschen Geographentages*, Vol. 34 (1965), 92–93.

through better living conditions and the second, a matching reduction in the birth rate through the new material and economic opportunities arising in the cities. Both men thus reasoned that once this stage of civilization was reached in other European and European-settled areas, a similar balance between the population and the physical and social environment could be reasonably expected. The correlation between population stress and technological level appears remarkably close in many countries, and not just those settled by Europeans, e.g., Japan.[4]

Yet the theory has not escaped criticism. As Lewis points out, we still do not really know why birth rates behave as they do, at least in the same way that we are familiar with reasons for variations in death rates. Why Americans, for example, should suddenly look with favor on larger families in the 1940's and 1950's, when before they were the subject of ridicule, still has not been satisfactorily explained. Most questioned, however, has been the implication of the theory that massive industrial development is the key to manageable population levels. Geographers have been especially critical of this suggested remedy for countries that have little or only meager amounts of the raw materials and fuels needed for intensive industrialization. Nor do they, or a growing number of scholars from other fields, see much hope of the proposed final stage of population equilibrium being obtained in many areas before widespread starvation is encountered, particularly where high birth rates combine with already large populations and industry still comprises but a small part of the economy.[5] Students of Marxian conviction, contrarily, foresee no problem if the inadequacies of distribution which they ascribe to the capitalist system are eliminated, and they reject the assumption inherent in the population cycle theory that industrialization eventually reduces population growth.

Cropland Expansion and Yield Intensification

A frequent corollary of attempts to estimate population capacities has been an evaluation of possible new areas for agricultural expansion. The humid tropics, with their generally abundant moisture and long growing seasons, have naturally received the most attention, although not always a favorable judgment. Earlier opinions were usually optimistic, writers like Ravenstein and Baker seeing the tropics as the great granaries of the future, and others, like Penck, Ballod, and Fischer, expecting them to become areas of the greatest population agglomerations as well. Later, when such researchers as Sapper and Ackermann produced evi-

[4] E. Wagemann, *Menschenzahl und Völkerschicksal* (Hamburg: Krüger, 1948); F. W. Notestein, "The Population of the World in the Year 2000," *Journal of the American Statistical Association*, New Series, Vol. 45 (1950), 335–45; W. S. Thompson, "Population: Is civilization endangered by the probability of a rapid Malthusian growth of the peoples in Asia and Latin America during the next five decades?" *Scientific American*, Vol. 182 (1950), 11–15.

[5] C. L. White and D. J. Alderson, "Industrialization: Panacea for Latin America?" *Journal of Geography*, Vol. 56 (1957), 325-32; W. Zelinsky, "The Population Prospect in Monsoon Asia," *ibid.*, Vol. 60 (1961), 312–21; T. Burke, "Food and Population, Time and Space: Formulating the Problem," *ibid.*, Vol. 65 (1966), 58–66.

dence that soils in many parts of the tropics were of low fertility, views of potential production in the low latitudes became more pessimistic. Whether newcomers to these zones could acclimatize themselves also became a hotly debated subject.[6]

Within the last two decades, however, a more optimistic trend has again developed. Food deficiencies are now being ascribed more to the lack of rational and intensive farming practices than to soil deficiencies. Gourou observed a surprising lack of agreement between population concentrations in tropical Africa and the quality of the soil. Even a negative correlation has become evident, many tropical cultivators avoiding the better-endowed valley bottoms because their farming techniques are easier to apply to the poorer soils of the terraces and plateaus. A growing number of geographers and soil scientists have also been revising their estimates of arable acreage in the tropics drastically upward. Hanson was already raising the question in 1945 of whether too sweeping generalizations had been made about the extent of good land, and by 1955 Gourou felt confident enough to assert that 50 per cent of Africa was cultivable.[7]

A mass of scientific research during World War II also appears to have discredited the common belief that most whites cannot acclimatize themselves to the tropics, thus reinforcing the argument of those who view the problems of adjustment as principally psychological, and not physiological. Thus, in the vital matter of mental activity Lee now maintains that it is the lack of motivation, and not the severity of tropical conditions, that is the primary disturber of mental processes. Expanding tropical studies, as James emphasizes, are beginning to reveal a much more varied climatic picture than heretofore assumed, and have led Stamp to conjecture that certain climates within the tropics may in fact be "nearest to the ideal for the physical and mental development of man and his survival." [8]

Unfortunately, the overwhelming share of acclimatization studies has been concerned with the white race; little has been written on the pioneering of tropical peoples or the mid-latitude Asians, such as the Chinese and Japanese. In light of the increasing tendency to attribute the

[6] Scharlau provides a good review of the optimistic attitudes toward the Tropics, *op. cit.*, p. 239; Baker's view is found in *op. cit.*, p. 25; Sapper, cited in Scharlau, *op. cit.*, p. 241; E. A. Ackerman, "The Geographical Meaning of Ecological Land Use," *Journal of Soil and Water Conservation*, Vol. 1 (1946) 63–66. The debate over acclimatization reached a peak at the International Geographical Congress in Amsterdam in 1938: *Comptes rendus du congrès international de géographie Amsterdam 1938*, Vol. 2, Sec. 3c: *Géographie coloniale* (Leiden: Brill, 1948), pp. 3–364.

[7] P. Gourou, "The Quality of Land Use of Tropical Cultivators," in *Man's Role in Changing the Face of the Earth*, p. 337; E. P. Hanson, *New Worlds Emerging* (London: Gollancz, 1950), pp. 131–38; P. Gourou, "Les conditions du développement de l'Afrique tropicale," *Genève-Afrique, Acta Africana*, Vol. 1 (1962), 49 f.

[8] D. H. K. Lee, *Climate and Economic Development in the Tropics* (New York: Harper, 1957), p. 100; P. E. James, "Man-Land Relations in the Caribbean Area," in *Caribbean Studies: A Symposium*, ed. V. Rubin (Seattle: Univ. of Washington, 1960), p. 15; L. D. Stamp, *The Geography of Life and Death* (London: Collins, The Fontana Library, 1964), p. 75.

white man's inability to settle the tropics in large numbers to the clash between his economic status and culture and those of the inhabitant native to the area, it would seem that further research with people less favored economically and perhaps also more compatible culturally might show greater potential settlement. Even such findings may not necessarily prove a solution to the overcrowded lands, so long as exclusionist immigration policies reign. The sharply critical view of such tactics held by the Indian economist, Mukerjee, and their equally vigorous defense, by Vogt, as a means of preserving living standards in the less-peopled countries would seem to indicate that the ultimate solution of the world's food problem lies more in international diplomacy than in the massive extension of agriculture into new areas.[9]

There is another school of thought, which is that the freezing of most of the world's population in its present location might not necessarily be a problem after all. With the best lands now mostly taken up, so goes this reasoning, it is only common sense to increase production in the established agricultural areas, where technological advances are most assured of successful application and extensive results. It is a view that has been, at least in the English literature, regularly advocated over the last three decades, although not all proponents have agreed on what developed area offers the greatest opportunities for yield intensification. This view has also received increasing support from accounts, like those by Harris of huge acre-yield increases and by Fels and others of presumably extensive acreage available for irrigation.[10]

The cartographic presentation by Visher of the potentialities of the established agricultural realm contrasts sharply with the map prepared by Hollstein, who sees the greatest possibilities for food production as being in the tropics (Figs. 10 and 11). Visher based his estimates of production favorability on the sums of ratings, ranging from 0 (practically no potential agriculture) to 4 (great potential agriculture) and applied to several physical and economic regional criteria. Hollstein represented potential production by the number of people that could be supported on one hundred hectares of land, a figure obtained by dividing a daily individual requirement of 2,500 calories into the total potential caloric production of the principal grain crop in an area. Hollstein and Visher even differ on the value of the tropical climate. The principal advantage of the tropics, to Hollstein, is the climate, but

[9] The views of Mukerjee and Vogt are contrasted in more detail by K. A. Sedlmeyer, *Geographie des Hungers* (Hagen, Westfalen: Pick-Verlag, 1962), p. 29.

[10] C. O. Sauer was one of the first who professed to see more of a future in technological advances in established production areas than in developing new lands. See his "The Prospect for Redistribution of Population," in *Limits of Land Settlement*, ed. I. Bowman (New York: Council on Foreign Relations, 1937), pp. 7–24. G. L. Mehren has continued the argument in "Geography, Geopolitics and World Nutrition," in *Food and Civilization*, ed. S. M. Farber (Springfield: Thomas, 1966), p. 122. See also C. D. Harris, "Agricultural Production in the United States: The Past Fifty Years and the Next," *Geographical Review*, Vol. 47 (1957), 175–93; and E. Fels, "Die Bewässerungsfläche der Erde," *Wiener geographische Schriften* (Festschrift Leopold G. Scheidl zum 60. Geburtstag, eds. H. Baumgartner *et al.*, Part I), Nos. 18–23 (1965), pp. 33–50.

it is criticized by Visher as being too warm for most of the world's more valuable crops and farm animals. Chang has recently made similar criticisms, asserting that the combination of much attenuated solar radiation, persistently high night temperatures, lack of seasonality, and excessive rainfall reduce "potential photosynthesis" and limit the possibilities of diversified agriculture.[11]

Continuing disagreements like these have brought a growing demand for more intensive inventories of land. Inventorying by countries of their land resources is now a widespread activity (see Chaps. 2 and 8), but not until 1949 was a step taken toward an international mapping program. The International Geographical Union at its sixteenth congress in Lisbon, at the request of Van Valkenburg, appointed a commission to study the possibility of a World Land Use Survey. More than sixty geographers of different countries are now working on the Survey, compiling a world land use map at a scale of 1:1,000,000. Mapping at such small scales, however, can become self-defeating: too much detail may be eliminated, as Reeds shows in his critique of Canadian land use survey sheets. Kostrowicki and his co-workers in Poland are one of the few national groups now making an attempt to fit their more detailed land use sheets within the broader classification headings being used by the World Land Use Survey.[12]

Comprehensive detail is also being sought in surveys of that part of agriculture which is less expressed in the landscape and more in terms of human conditions and problems. In such a study of rural Malaya, Gosling has found a variety of indexes helpful, such as crop production per agricultural worker, percentage of self-sufficiency in food crop production, per capita income, and percentage of land in cash agriculture. How valuable the finding of relationships of this kind can be, once enough statistics are available, is shown in the several worldwide agricultural maps by Ginsburg, in which countries are ranked according to the world mean.[13]

[11] S. S. Visher, "Comparative Agricultural Potentials of the World's Regions," *Economic Geography,* Vol. 31 (1955), p. 85, Fig. 2; Hollstein, *op. cit.,* backpaper. Visher's pattern, not surprisingly, differs comparatively little from Zelinsky's, in which current technological progress is regionalized on the basis of a "population–resource ratio." Cf. W. Zelinsky, *A Prologue to Population Geography* (Englewood Cliffs, N.J.: Prentice-Hall, 1966), pp. 108–109. C. Clark has presented much the same pattern as Hollstein, although his bases for computation are coarser, not having taken into account either the amount of arable land available or physiological needs. His maps appear in *Population Growth and Land Use* (New York: St. Martin's, 1967), pp. 145–48. J. H. Chang's criticism appears in "The Agricultural Potential of the Humid Tropics," *Geographical Review,* Vol. 58 (1968), 333–61.

[12] S. Van Valkenburg, "The World Land Use Survey," *Economic Geography,* Vol. 26 (1950), 1–5; L. G. Reeds, "The Land-Use Survey of the Niagara Peninsula: A Critical View," *Geographical Bulletin,* Vol. 7 (1965), 203–11; J. Kostrowicki, *The Polish Detailed Survey of Land Utilization: Methods and Techniques of Research,* Dokumentacja Geograficzna, Zeszyt 2 (Warszawa: Instytut Geografii, Polska Akademia Nauk, 1964).

[13] L. A. P. Gosling, "The Location of 'Problem' Areas in Rural Malaya," in *Essays on Geography and Economic Development,* ed. N. Ginsburg (Chicago: Univ. of Chicago, 1960), pp. 124–42; N. Ginsburg, *Atlas of Economic Development* (Chicago: Univ. of Chicago, 1961), pp. 46–55.

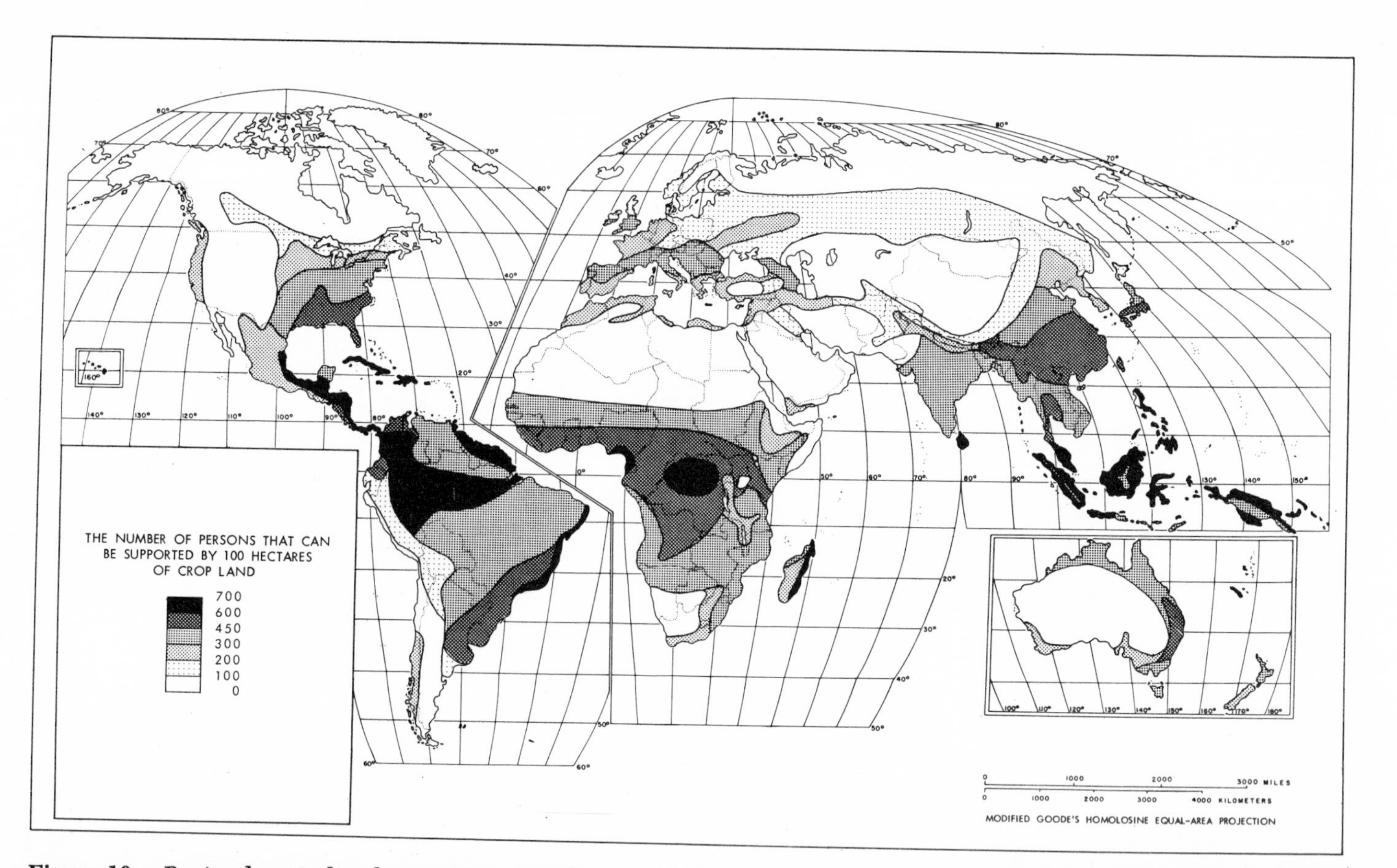

Figure 10. *Regional agricultural capacities, according to Hollstein. [By permission of the author and VEB Hermann Haack, Geographisch-Kartographische Anstalt Gotha/Leipzig.]*

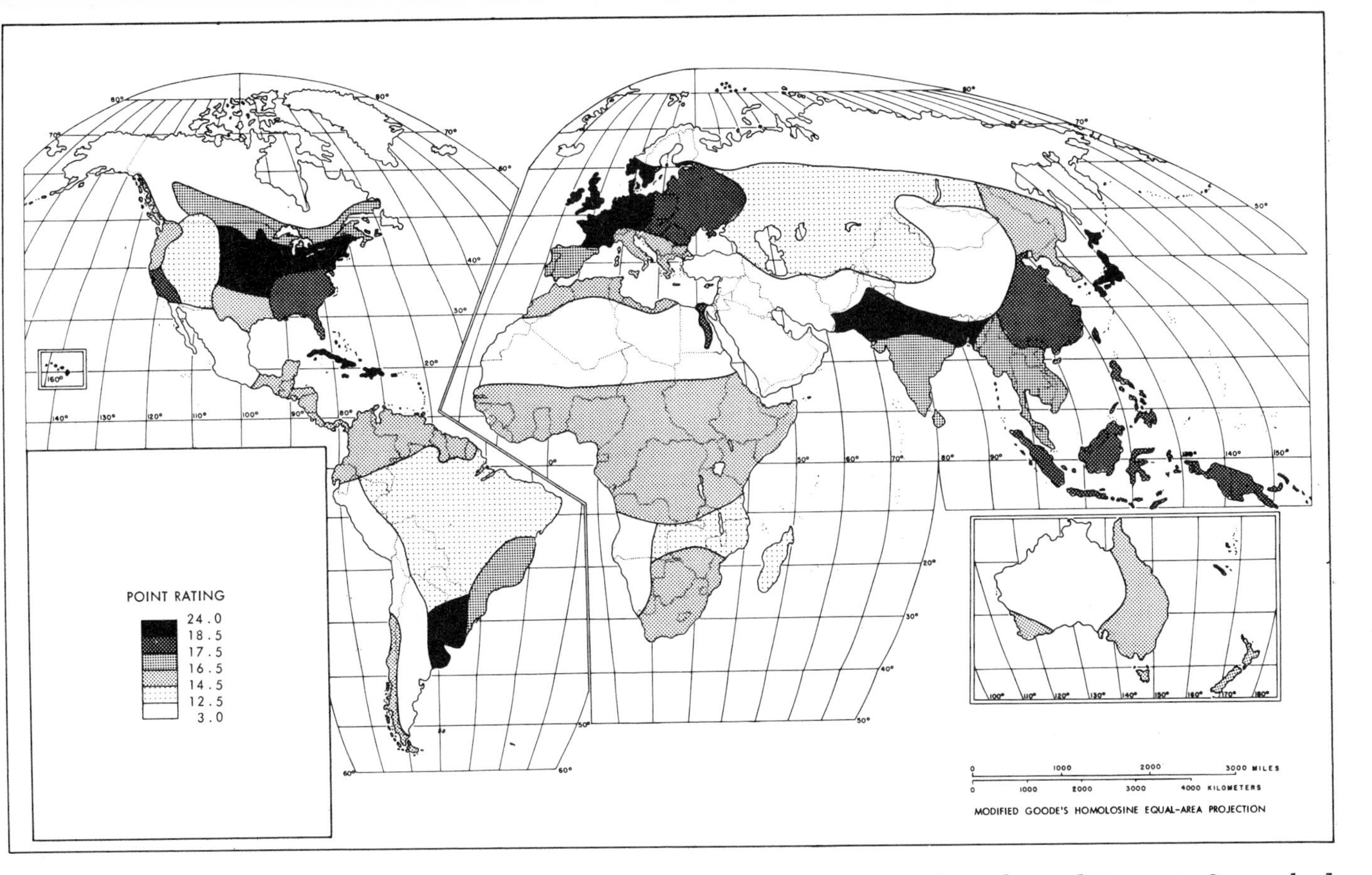

Figure 11. *Regional agricultural capacities, according to Visher. [By permission of the author and* Economic Geography.*]*

Dietary Deficiencies

Increasing concern with more regional detail also has been evident in the search for information about the specific kinds of dietary deficiencies suffered by man. Although this activity began much later than work on the more general and quantitative aspects of food sufficiency, interest was high enough as far back as 1919 to warrant a call for insertion in the university curriculum of a "geography of nutrition." [14] And by 1928, an investigation of world dietary conditons had been inaugurated by the Health Organization of the League of Nations.

These and later attempts to deduce the distribution of dietary quality placed special emphasis on the areal variation in caloric consumption. It was also the base used for most of the studies on population capacity, like Hollstein's, which were related to physiological needs. It is still a widely accepted index of dietary quality, although arguments persist over just how widespread undernourishment, i.e., the insufficient intake of energy-supplying calories, is. The most contradictory views are probably those of the Food and Agricultural Organization (FAO), the successor to the Health Organization of the League of Nations, and Bennett. All of Africa, Asia, and Latin America (except Uruguay and Argentina) are afflicted with serious undernourishment, according to the FAO, while Bennett argues that the organization underestimates caloric intake and overestimates caloric requirements in underdeveloped areas. In fact, Bennett believes, were enough allowance made for differences in body size and activity, and an accurate account taken of climatic conditions and age and sex distribution, the information would show few caloric deficiencies anywhere. Unfortunately, as Blynn points out so well, it is in the very countries whose populations are most seriously affected that direct evidence of nutritional intake is still most lacking.[15] Much of the classification of undernourishment is based on indirect evidence obtained by examining one ratio, the proportion of calories obtained from starchy staples, presumed to be low in health-protecting nutrients, in the total calorie intake. If the ratio is at least two-thirds, inference of malnutrition appears safe, although even this evidence cannot be considered complete.

The attempt to obtain an ever more comprehensive view of variations in dietary quality, both areally and qualitatively, has gained momentum since 1940. Interest then began to be focused on an additional aspect of malnutrition, malnourishment, or the insufficient intake of health-protecting vitamins, minerals, and protein. This wider definition of the problem can be perceived in such works as Huntington's comparisons of national diets with respect to the relative share of "non-

[14] A. Lipschütz, "Die Aufgaben einer Ernährungsgeographie und ihre Anteil an einer vergleichenden Ernährungslehre," *Geographische Zeitschrift*, Vol. 25 (1919), 137–56.

[15] M. K. Bennett, *The World's Food* (New York: Harper, 1954), p. 200; G. Blyn, "Controversial Views on the Geography of Nutrition," *Economic Geography*, Vol. 37 (1961), 74.

protective" and "protective" foods. Even more instructive are the world-wide maps, by May and by Cépède and Langellé, of several dietary characteristics. May's maps contain especially detailed areal delineations and feature three dietary distribution patterns, one of areas whose diets are adequate in both "protective" and "energy" values and the other two of areas where diets lack either just "protective" value or both "protective" and "energy" value. Regions of specific vitamin deficiencies in Southeast Asia, Africa, and Latin America are also mapped. Although the areal patterns of undernourishment and malnourishment vary from one map to another, depending on the author, there seems to be general agreement that malnourishment is more widespread than undernourishment and that the two scourges together clearly outline the underdeveloped areas, particularly Southeast Asia. Burke's recent maps of nutritional adequacy, based on FAO statistics, add further backing to this conclusion.[16]

The possibility of another comprehensive view of what constitutes dietary quality, one that allows the areal variations to be represented by a single index, has been advanced by Zobler. Using the United States and Japan as examples, he derives the "nutrient production relative," or total production of a given nutrient from all the foods that make up the diets of the two countries. This index is obtained by first multiplying the "area relative" (the ratio of the area producing the foods containing that nutrient to the total food producing area) by the "yield relative" (the ratio of the yield per hectare of the foods producing that nutrient to the amount needed to meet the recommended allowance of food constituents for 2.5 persons per year). The resulting product is then multiplied by the relative content of each food in that nutrient. The entire procedure is carried out for each of ten basic nutrients, two of which are calories and proteins, the first expressed in number and the second in percentage. The sum of the computations is the "multiple nutrient production relative," a single value that expresses the "general nutrient balance" of the food economy of each country.[17]

As evaluations and mapping of dietary quality have become more detailed, however, the assumptions and criteria on which such work is based have come more into question. Nutritionists now feel that the areal variation of protein quality is of more dietary significance than protein quantity, the kinds of amino-acid combinations being the crux of the problem. And, as Fonaroff observes in his attack on Huntington's earlier quantitative emphasis, low-quality protein can never be compensated for by increased intake. Unfortunately, protein quality is still not adequately

[16] E. Huntington, "The Geography of Human Productivity," *Annals of the Association of American Geographers,* Vol. 33 (1943), 13–19; J. M. May, "The Mapping of Human Starvation: Diets and Diseases," *Geographical Review,* Vol. 43 (1953), 403 f.; *idem, Atlas of Diseases: Study in Human Starvation, 2. Diets and Deficiency Diseases* Pl. 9; M. Cépède and M. Langellé, *Economie alimentaire du globe* (Paris: Médicis, 1953); T. Burke, "Food and Population, Time and Space: Formulating the Problem," *Journal of Geography,* Vol. 65 (1966), p. 61, Fig. 3.

[17] L. Zobler, "A New Areal Measure of Food Production Efficiency," *Geographical Review,* Vol. 51 (1961), 549–69.

presented cartographically. We also now know that metabolic levels vary far more areally than the figures set up as "recommended minimum nutritional requirements" would lead one to believe. Yet these variations apparently are fully compatible with maximum health, the body reacting to a reduction in the recommended nutritional levels by merely establishing a new level of equilibrium, but with less reserves.[18]

Ascertaining the principal reasons for the many differences in dietary quality over the earth has been more difficult than explaining the situation in food production, where the physical environment plays such an important role. Certainly, tastes in food have strong roots in local cultures, and although amazingly few students have seriously concerned themselves with this connection from a geographic point of view, Sorre's work on a "geography of dietary regimes" is superlative. An excellent later contribution was the consideration given by Simoons to the role of taboos in Old World diets. The physical reasons that have been given for variations in dietary quality are naturally many, perhaps the most popular being soil impoverishment. May conceives this destructive process as a vicious circle in which inability to purchase commercial fertilizers or even to add enough human and animal manures to the soil (common in starvation areas) leads to a multiplying impoverishment of both soil and man.[19]

Economic and social circumstances have been most studied by those seeking the major causes of undernourishment and malnourishment. The commercialization of basic foods or the substitution of less nutritive foods, the use of the best land for commercial crops, inferior farming methods and implements, exorbitant land rates, and lack of coordination between supply and demand have all been condemned as obstacles to improvement of the food supply. But which of these barriers is the most crucial is still moot. Castro is convinced that European and American colonialism has been responsible for most of the problems, although Sorre has rightfully noted that many of them existed before foreign groups arrived.[20]

Disagreements within the economic and social context also include reservations about some of the most popular views of the relationship between population growth and food supply. The prevailing belief, particularly among supporters of Notestein and Thompson's population cycle theory, that the raising of income levels will inevitably improve dietary conditions, has evoked a strong qualifier from Bennett. He believes that ignorance and food preferences can, at any one time, influence nutritional quality at least as much as poverty does. Just as popular an opinion, that

[18] L. S. Fonaroff, "Was Huntington Right about Human Nutrition?" *Annals of the Association of American Geographers*, Vol. 55 (1965), 365–76.

[19] M. Sorre, *Les fondements biologiques*, Vol. 1: *Les fondements de la géographie humaine* (Paris: Armand Colin, 1951), pp. 247–90; F. J. Simoons, *Eat Not This Flesh* (Madison: Univ. of Wisconsin, 1961); May, *The Mapping of Human Starvation*, pp. 403 f.

[20] J. de Castro, *Géopolitique de la faim*, 2d ed. (Paris: Editions ouvrières, 1962); M. Sorre, "La géographie de l'alimentation," *Annales de géographie*, Vol. 61 (1952), 199.

too rapid a rate of population growth can lead to dietary degradation, has incited even more debate. "Malthusianists" view this sequence as inevitable, while Marxists believe it illusory. Adherents of the population cycle theory adopt somewhat of a middle ground, accepting the possibility of a damaging disparity between the two variables, but only so long as technological development has not been able to achieve equilibrium. The most revolutionary of the more recent views is that of Castro, who contradicts both the Malthusianists and the Marxists, claiming a close cause-and-effect relationship but one that makes dietary quality the determinant of population growth rather than the reverse. It is an opinion that has been hotly attacked from many quarters; it is also, however, powerful evidence that researchers continue to seek all possible answers to man's most vital problem, his ability to feed himself.[21]

[21] Bennett, *op. cit.*, pp. 208, 222–26; Castro, *op. cit.*, pp. 105–10.

ıdex

ANDONMENT(S), 44, 106, 108, 142, 146
sentee-urban owners, 79
climatization, 155
kerman, E. A., 14, 33, 34n, 154, 155n
re-yields, 22, 152
reage:
available for irrigation, 156
controls, 86
histories, 86
restrictions, 86
dis Ababa, 63
ministrators, 81
rial photo coverage, 23
rial photograph(s), 18, 23, 36
in field work, 20
of infrared radiation, 20
rial photography, 19–20
forestation, 46
rica, 83, 147, 160
climatic cycles in North, 143
East, 51
export commodities in, 112
population concentrations in tropical, 155
tribal areas of, 81
vitamin deficiencies in, 161
West, 84
gglomerated settlement(s), 98, 99, 105, 108
grarian:
geography, 6, 7
structure, 11, 121–22
gricultural:
areas of the Old World, 107
boundaries, 137
domains, 9
enterprises, 27
foreposts, 134
forms, 10, 99
fringe, 88
geographers in the United States, 22
geography, 25, 79, 103, 120
advocacy as a field, 15
as an agricultural science, 4–5
American, 21
applied aspect, 2
archival research, 20
as areal variation of agriculture, 2
as a bridge field, 4, 6–7
concern with the physical environment, 31
as description, 1
as distinct from economics, 3
as distinct from geography, 4, 6
ecological, 6
generalizations, 21
greater independence of, 4
historical aspects of, 99, 101
human side of, 44
mapping of small areas, 17
mathematically-based research in, 29
in the Middle Ages, 1
physiognomic, 6
practical objectives, 14–15
"pure" objectives, 14–15
readings in, 128
regional, 128
regional studies in American, 121
as relationships, 9–11
religious motives in, 76
research themes in, 37
statistical, 6
in the "system of agricultural sciences," 5
themes in, 151
in the university curriculum, 15
islands, 132
land, 44, 62, 116–17, 144–46, 152–53
landscape(s), 6, 10, 57, 78, 94, 97, 107, 109, 112, 120, 139, 145
maps, 24, 157
outliers, 133
patterns, 34, 36
political-, relationships, 85
population(s), 40, 119
prices, 148
production, 39
region(s), 114, 118, 126–27 (fig.). See *also* Areas; Regions
American, 136
in California, 22
comparison of total, 129
crop-combination, 27
cultural aspects, 18
"cultural origin, maturity, and change" of the, 129
delimiting of, 136
economic aspects, 18
and the fractional code, 18
functional, 119
German, 129
as the highest goal of study, 10–11
"historical-social," 24
livestock-combination, 27
major, 123
more culturally oriented, 122
in Ontario, 35
over-all, 120
total, 121, 124, 128
in the Western United States, 100
Whittlesey's worldwide map of, 24
world map of, 124
regionalization, 27, 117
resources, 143, 147
sciences, 2, 4–5, 32
settlements, 130
space, 9
sufficiency, 119

Agriculture (*cont.*)
surpluses, 148
systems, 13, 20
technology, 5, 43, 55
terrain-, relationships, 36
"topographers," 1
tradition, 24
workers, 83, 157
zoning, 145
Agricultural Atlas of the U.S.S.R., 92
"Agricultural formation," 5, 10, 120
"Agricultural ladder," 80
"Agricultural man," 45
"Agricultural Productivity of the World's Soils," 56
"Agricultural types," 124
"Agricultural zones," 111, 116–17
Agriculture, 132, 145, 157
altitudinal boundary of, 131
ancient, 101
antiquity of local, 95
commercialized, 123
competitors of, for space, 146
European, 59
extension of, into newer areas, 156
great potential, 156
land in cash, 157
margins of, 130
Mediterranean, 123
a more industrialized, 148
no potential, 156
North American, 117
plantation, 100
polar limits of, 130–31
possibilities of diversified, 157
potential limits of dry-land, 130
regional studies of North American, 117
retreat of, 136
site of the first, 94
social aspects of, 22
stages in, 100
subsistence, 82
terrain available to both cities and, 144
theoretical aspects of, 59
transformation of, by government, 92
typology of, 125
in underdeveloped areas, 148
Welsh, 101
Agriculturists, 100. *See also* Farmers; Peasants; Planters; Renters; Rural man
"Agroclimatic belts," 33
Ahnert, F., 140
Air pollution, 146, 147n
Air temperature, 33
Albedo, 45
Alberta, 89
Albrecht, W. A., 46n, 140, 147
Albright, E. D., 46n
Alderson, D. J., 154n
Alexander, J. W., 63
Alfalfa areas, 66
Alisov, B. P., 45, 46n
Allegheny Highlands, 44
Alluvial fans, 35
Alps, 35
Alsace, 49, 105
Altitude, 115, 131
Altitude boundaries, 130–31, 135 (fig.)
Amenities, 134
American(s):
agricultural geography, 121
agriculture, 21
colonialism, 162
dietary demands of, 151
farmers, 76
of Japanese ancestry, 74
population, 42
regionalization of, farming systems, 117
tobacco farms, 112
Amino-acid combinations, 161
Amish, Old Order, 77
Amortization costs, 44
Andes, 50, 132
Andreae, B., 22n, 44n, 60, 65, 115, 117–18, 142
"Anecumene," 133
Anglo-America, 82
Animal(s), 97. *See also* Cattle; Livestock; Sheep
-crop combinations, 117
distribution maps of major crops and, 111
farm, 157
foods, 78
genetics, 34
geography, 34
husbandry, 40, 108, 123
migration or transport of, 41
optimum soil area for, 36
power, 44
regionalization of crops and, 111
relationships between domesticated plants and, 12
routes followed by plants and, 95
stall-fed, 41
"Annual frequency of cultivation," 38
Anuchin, V. A., 124, 125n
Apperception tests, 149
Apple economy, 35
Arable land(s), 89, 128, 131, 149, 151–52
Archeological evidence, 95
Archeological research, 95
Archival material, 22–24
Archival research, 20–21
Arctic Front, 31
Area(s) 123–24, 128. *See also* Agricultural, regions; Regions
agricultural, of the Old World, 107
"agricultural-autarkic," 119
"agricultural-deficiency," 119
"agricultural-surplus," 119
alfalfa, 66
centrally oriented agricultural, 119
of "chronic crisis," 147
compact agricultural, 135 (fig.)
considered transitional, 136
core, 137
"deficient" food production, 153
established agricultural, 156
farming, 136
fragmented agricultural, 135 (fig.)
fruit, 144
inhabited, 135 (fig.)
one-crop, 144
Peace River, 63, 132–33
positioning of sample, 19
ranching, 79
Rio de Janeiro-Sao Paulo, 56
rural, 148
San Diego, 40
San Francisco Bay, 40, 119
service, 120
Slavic, 98
specialty-crop, 146
of stress, 153
"surplus" food production, 153
tribal, 81
underdeveloped, 112, 148, 160–61
"Area relative," 161
Areal:
associations, 10
differentiation, 2
specialization, 43
variation(s), 2, 4, 11, 111, 151, 161
Argentina, 160
Arid regions, 146
Aridity, 88, 149
Arizona, 67, 134
Arroyo cutting, 143
Asia, 160
Southeast, 51, 56, 77, 99
archeological research in, 95
farmers in, 81
hill and mountain lands of tropical,
intensively farmed regions of, 90
labor intensity in, 63
tribal areas of, 81
vitamin deficiencies in, 161
southeastern:
British in, 84–85
Dutch in, 84–85
plains in, 34
Southwest, 99
climatic cycles in, 143
grain farmers in, 95
Asians, 155
Athens, Georgia, 19
Atlantic, 103
Atlantic Seaboard, 39
Atlas of Britain and Northern Ireland, 2
Atlases, 26, 92, 123
Atmosphere, 46–47
Atmospheric modifications, 46
Atomic explosion, 140
"Atomization," 144
Aufrere, L., 53n, 78n
Augelli, J. P., 74, 76, 80, 81n, 89
Austin, M. E., 114n, 140n
Australia, 42, 56, 82, 90
Avon Valley, 79

"BACK FURROW," 49
Bahia, 87
Baker, O. E., 21–22, 42–44, 65, 106–10 117, 120, 123n, 136, 148, 15 153n, 154, 155n
Balkan countries, 102
"Balks," 50
Ballod, K., 151, 152n, 154
Barley, 31, 36
Barn(s). *See also* Buildings; Farmsteads
cheese factories converted into, 106
combined houses and, 107
New England connecting, 99
Pennsylvania, 99
quality of, 74
relocating, 66
Barnes, C. P., 90n
Barnes, F. A., 125n
Barnes, J. A., 27

rrows, H. H., 14
rton, T. F., 31, 32n
rtz, F., 90, 122
tes, M., 142
umgartner, H., 86n, 100n, 156n
ard, C. N., 35
havioral sciences, 13
lgium, 91
ll, T. A., 105
nefit-cost analysis, 145
ngal, West, 31
nnett, H. H., 53, 139, 140n, 141n
nnett, M. K., 31, 32n, 113, 131n, 160, 162, 163n
resford, M. W., 49
rnhard, H., 2, 4–5, 14
st, R. H., 67
uermann, A., 41, 42n
atia, S. S., 21, 22n
hl, M., 39
ch, J. W., 19, 62, 63n, 150
d, E. T., 86, 87n, 122n
th rates, 148, 153–54
ache, J., 44
aut, J. M., 19, 150
e Mountains, 150
vn, G., 25, 160
al, F. W., 87
bek, H., 143
dy activity, 160
dy size, 160
okkeeping, 44
rder zone, 91
rn, M., 132, 133n
serup, E., 148, 149n
undary(ies), 27, 125, 137
"absolute" crop, 33
advancement of the cropping, 130
agricultural, 132, 137
altitude, 130–31, 135 (fig.)
in the Canadian northlands, 130
controversies, 89
crop, 25, 42, 130, 131
of different field combinations, 115
dry, 130
ecological, of crops, 33
economic, of crops, 33
farming, 131
forest, 131–32, 135 (fig.)
of irrigated land, 130
of permanent settlements, 135 (fig.)
polar, 130
polar agricultural, 130–31
proposals, 133
of the region, 120
regional, 28
reporting-area, 118
sciences, 7
of shifting animal husbandry, 41
snow, 135 (fig.)
studies, 136
values, 32
where a, lies, 136
zone, 91
wden, L. W., 29n, 109
wen, W. A., 47, 48n
wman, I., 11–12, 82, 133, 156n
acey, H. E., 120, 128, 129n
aidwood, R. J., 94, 143
azil:
Bank of, 87
failure of Germans in southern, 74
foreign groups in, 73
Japanese colonies in, 76
success of the Dutch in southern, 74
Brinkman, L. W., 28n
Brinkmann, T., 59, 67, 69n, 117
Britain, 20, 145
agricultural results obtained in, 151
correlation between slope and land use in, 36
scholars of, 123
sheep in, 87, 102
British:
Agricultural Returns, 20
Columbia, 35, 89, 104, 136
Land Survey Office, 19
in Southeastern Asia, 84–85
Brooder houses, 106
Brown, R. H., 25, 26n, 74, 75n, 101
Brown, S. E., 63n
Brunet, P., 105, 122
Brush, J. E., 120, 128, 129n
Bryson, R. A., 28
Buchanan, K., 59, 60n, 79
Buchanan, R. O., 8, 11, 137
Buchholz, E., 131
Buenos Aires, 60, 91
Buildings, 108, 112. *See also* Barns; Farmsteads
abandonments of, 106
better farm, 72
farmstead, 107
high cost of land and, 104
housing married and single workers, 81
specialized, 106
Bunge, W. B., 19, 27
Bureau of Agricultural Economics, 117
Burger, G., 34
Burke, T., 154n, 161
Burley, T. M., 122
Burton, I., 36, 66, 149, 150n
Busch, E., 117
Butter, 65
Butzer, K. W., 143

CADASTER RECORDS, 105
Caesar, 1
California, 39, 44, 54, 104, 124, 144
agriculture, 101
air pollution in, 147n
alluvial fans in, 35
Coast Ranges, 35
county groupings, 119
dairy farmers in southern, 61
dairy farmsteads in, 67
dairying in, 73
deserts of southeastern, 132, 134
export-oriented agriculture of, 67
farmers, 150
irrigation in the San Joaquin Valley of, 80
Italians in dairying in, 73
Japanese in, agriculture, 73
orange groves in southern, 144
southern, 19, 26, 40, 101
Swiss dairymen in, 76
urban centers, 144
valleys, 37
"Calm," periods of, 103
Caloric:
consumption, 160
deficiencies, 160
production, 113, 153, 156
requirements, 160
Calories, 156, 160–61
Canada, 46, 72–73, 76, 123, 128
fruit regions in, 115
major fruit area of, 144
polar agricultural margins in, 132
recent settlements in Western, 89
regional specialization in, 64
relationships between the United States and, 85
Canadian:
farmers, 132
northlands, 130
plains, 73
prairies, 77
settlements, 91
Canals, 54
Capetown, 59
Capital, 75
intensity, 63
investment, 109
lack of, 149
Capitalist system, 154
Carbon dioxide, 36
Carol, H., 121, 124
Cars, 106
Cartograms, 24–25
Cartography, 24
Cascades, 135 (fig.)
Cash-grain farming, 28, 34, 36
Castes, 81
"Casting," 49
Castro, J. de, 36 37n, 162–63
Cattle, 60, 112. *See also* Animals; Livestock
Celts, 98
Census:
of agriculture, 112
Bureau, 21, 111, 113
data, 23
definitions, 22–23
information, 20
reports, 23
statistics, 118
U. S. Agricultural, 39
Central places, 129
Centrality, 67
Cépede, M., 161
Cereals, 9, 59, 102, 115
Ceylon, 51, 100
Chang, J. H., 33n, 157
Chang, S. D., 93n
Change, 101
Chapman, L. J., 140
Chardon, R. E., 80, 81n, 88, 105
Chatham, R. L., 145
Cheese, 65
factories, 106
foreign, industry, 73
"national types," 74
regions, 122
Chemistry, 152
Chi-square test, 28
Chicago, 60, 64
Chicken houses, 106
Childe, V. G., 94
Children, 148

China, 43, 50–51, 81, 92
Chinese, 155
Chisholm, M., 59, 61n, 65–66, 88, 108
Cholley, A., 10–11, 121
Christian calendar, 76
Chung, Y. P., 93n
Citrus:
  acreage, 33
  extension of, over the Mediterranean area, 78
City(ies), 148
  as centers of culture and economic change, 146
  gentle terrain available to agriculture and, 144
  land use zonation near, 67
  largest, in the Eastern United States, 146
  loss of farmland to expanding, 143
  migration of farmers into the, 147
  milk production in the vicinities of, 64
  North American, 145
  opportunities arising in the, 154
  rapid horizontal growth of, 144
  sprawl of, 145
Civilization, 148, 154
Clark, A. H., 27, 97, 142
Clark, C., 157n
Classification. *See also* Regionalization; Zonation
  broader, headings, 157
  Hahn's, 124
  regional, 123
  soil, system, 56
  Whittlesey's, 124
  worldwide, schemes, 122
Clawson, M., 146, 147n
Cleveland, 19
Clibbon, P. B., 136
Climate(s), 99, 121, 131
  of ancient Germany, 45
  classifying, 33
  dependence of the transport route on, 63
  effects on, of a forest cover, 45
  effects on livestock, 34
  human alterations of, 45
  influences of the removal of vegetation on, 45
  quality of, 152
  relationships between barley and, 31
  role of, 152
  "tender fruit," 144
  tropical, 155–56
Climatic:
  analogs, 32, 113
  change, 46
  classification, 32
  cycles, 142
  facts, 10
  fluctuations, 142
  homologs, 143
  regions, 118
  rhythm, 37
  systems, 113
  theories, 143
  variations, 31
Coal, 152
"Coconut landscape," 129
Coconut production, 34
Coefficient of correlation, 28
Cohen, S. B., 133–34n
Colby, C. C., 12, 14, 18, 19n, 35, 80, 121, 122n
Collecting, 100
Colombia, 89
Colonial policies, 84
Colonialism, 162
Colonies:
  absorption of Japanese, 76
  European, 73
  immigrant flow to established, 73
Colonization:
  agricultural, 88
  comparisons of, 128
  early, in Slovenia, 98
  program in the Dominican Republic, 89
  role of agriculture in, 101
  successful, 89
Colorado, 109
Colorado Front Range, 135 (fig.)
"Combination forms," 124
Commerce, 121
Commercial plantation crop tillage, 123
Commercialization, 162
Commodities, 112, 121
Commune, Chinese, 83
Communication, 109
Competition, 144–46
Competitive overlaps, 136
Computer, 118
Computerized research, 29
Conklin, H. C., 76, 77n
Conservation, 74, 142, 149
Conservative milieu, 9
Consolidation:
  farm, 80, 108
  farm-, programs, 103
  of fields, 105
  of smaller farms, 66, 87–88, 104
Controls, acreage, 149
Cooperative organization, 80
Coppock, J. T., 21, 26n, 27, 29, 117–18
Coquery, M., 79
Core areas, 23
Corn, 60, 102
  area in West Germany, 32
  Belt, 31, 64–65, 101, 129, 136
  -hog-beef cattle farming system, 64
  -Soy Belt, 101
Corrals, 61, 67
Correlative sciences, 7
Cotton, 111
  Belt, 43, 86, 101, 136
  in the Cotton Belt, 101
  farming, 43
  regions, 43
  retreat to the best lands, 43
  shift of, 86
  surpluses, 86
  in West Pakistan, 87
Counties, 22, 107–8
Courtenay, P. P., 84, 85n
Covariation, 28
Cox, H. J., 35
Cozzens, A. G., 73
Credner, W., 18, 124
Critchfield, H. J., 35
Crop(s), 18, 27–28, 31, 66, 120–21. *See also* Plants
  acreage, 22
  in Africa, 84
  area, 150
  basic, distributions, 112
  boundaries, 25, 42, 130–31
  calendars, 38
  cash-grain, 60
  choices, 9, 70
  climatically tolerant for the region,
  commercial, 162
  distribution, 21, 33
  distribution maps of major, 111
  "edaphic optima" for, 36
  in England and Wales, 65
  fodder, 102, 115, 117
  forage, 60
  forecasting yields of, 25
  grain, 111
  grown on plantations, 84
  growth, 33
  history of regional varieties of, 39
  importance of the key, 116
  landscapes, 38
  -livestock associations, 25
  maturity rates of specific varieties of,
  one-, areas, 144
  optimum areas for, 32
  optimum soil areas for, 32
  plantation, tillage of, 123
  of the planting type, 63
  potential, growth of, 113
  principal food, 153
  production per agricultural worker, 1
  reduction in acreage of major, 86
  regionalization of, 111
  regions, 2, 9
  root, 58 (fig.)
  rootstock, 39
  rotation, 9, 117
  shifts, 101
  specialty-, areas of the West, 22, 14
  staple subsistence, 112
  support, 148
  tonal contrasts in photographs, 20
  tree, 141
  truck, 39, 79
  world's more valuable, 157
"Crop weight," 117
Cropland:
  along a route, 63
  boundary between forest and, 131
  contraction, 43
  expansions of, 46
  in fodder crops, 115
Cropping, 9
  advancement of boundary of, 130
  calendar, 38
  operations, 41
  pattern, 37
  systems, 92
"Cropping economies," 115
"Cropping wheel," 25
Cultivation, 45, 65
Cultural:
  attitudes, 73
  features, 17
  landscape, 6
  legacy, 8
  motives, 99
  processes, 129
  traits, 98
"Cultural islands," 72

ılture(s):
European, 25
garden, 123
hoe, 99–100, 123
local, 162
plantation, 123
plow, 99, 123
rural, regions, 83
urry, L., 38
urtis, J. T., 45, 46n
urwen, E. C., 100n
ustoms, 103
ycle(s), 108
dry, 142
of soil destruction, 142
zechs, 72

AHLBERG, R. E., 35, 63n
airy:
Belt, 64–65
national groups in the American, 74–75
"new," 136
"old," 136
subregions, 136
farmsteads, 67
products, 66
regions, 128
airying, 59, 64, 75, 85, 109, 123
"drylot," 109
Hawaiian, industry, 74
Italians in, in California, 73
arby, H. C., 63, 64n
ata-processing equipment, 29
avidson, F. A., 34
avitaya, F. F., 141
Dead furrow," 49
easy, G. F., 34
eath rate(s), 154
curve, 140
impact of the industrial revolution on birth and, 153
ecision-making behavior, 70
effontaines, P., 38–39, 42, 46n, 76–77, 78n, 90, 99
eforestation, 45–46
Delamarre, M. J. B., 49
Deltas, 34
emangeon, A., 98, 99n
'Demodynamic alternation law," 153
Demographic rates, 80
Denevan, W. M., 143
'Denominator" elements, 18
Denudation rate, 140
Department of Agriculture, 32
Derruau, M., 50, 108
Description, 1
Desert(s), 9, 88, 132, 141
lands, 151
oases of the southeastern, 102
of southern California, Nevada, and Arizona, 134
Deshler, W., 112n
Deshpande, C. H., 81
Despois, J., 50
Dessication, 143
"Development stages," 142
Diagrams, 23–24
and agricultural maps, 26
block, 25
moisture-utilization, 25
"regional," 25
Dicken, S. N., 23, 24n, 25, 120, 121n, 136
Diet(s), 161
comparison of national, 160
effect of soil depletion on, 147
Old World, 162
western European, 9
Dietary:
conditions, 160, 162
deficiencies, 160
demands, 151
distribution patterns, 161
quality, 160–63
Diffusion:
of the decision to irrigate, 29
of the New England connecting barn, 99
of planting cultures, 95
of wells, 109
Dion, R., 101, 125
Dispersed settlement, 99
"Displacement energy," 107
Diversification, 80
Dix, E., 85n
Dobby, E. H. G., 34, 122
Dominica, 80
Dominican Republic, 89
Dozier, C. L., 132n
Drainage:
accelerated, through deforestation, 45
and ridged fields, 47, 50
Dresdov, O. A., 45, 46n
Drought(s), 45, 94, 142–43
frequency of, 149
perceptions of the hazard of, 149
records of, 149
Drumlins, 35, 140
Dry:
boundaries, 130
cycles, 142
lands, 88
advances of irrigation into the, 103
agricultural margins of the, 133
agriculture, 130
Soviet, 130
"Drylot dairying," 109
Dry matter-production probability regime, 38
Dumont, R., 81
Dunn, E. S., Jr., 13, 63
Durand, L., Jr., 24, 35, 62n, 64, 65n, 74–75, 85n, 86, 87n, 106, 119, 121, 122n, 136, 140, 144, 145n
Durban, 59
Dutch, 72, 74
farmers in southwestern Ontario, 76
in Southeastern Asia, 84–85

EASTERN EUROPEAN COUNTRIES, 85, 109. *See also* Europe
Ecker, L., 141n
Ecological agricultural geography, 6
Economic:
drives, 98
forces, 120, 131
form, 18
geography:
of agriculture, 6
cultural emphasis, 8–10
as distinct from economics, 3–4
environmentalist views, 7
objectives of, 2–4
possibilist concepts, 8
practical objectives, 14–16
"pure" objectives, 14–16
reciprocal concepts, 8
as regional geography, 14
as a science, 12
and social objectives, 14
texts, 123
theoretical, 12–13
traditional, 13
landscape, 10
level, 2
motives, 104
policies, 84
soil, 56
Economic and Social Atlas of Greece, 26
Economics, 3–4
"Edaphic optima," 36
Eddy, F. W., 143
Egypt, 76
Ehlers, E., 19, 25, 46n, 63, 64n, 73, 88–89, 128, 132–33
Eidt, R. C., 47, 48n, 89, 132
Ejidos, 105
Elliott, F. F., 117
Ely, R. T., 59
Emigration, 89
Encroachment:
selective, 144
urban, 144–46, 150
Energy balance, 45
Engelbrecht, T. H., 2n, 21, 22n, 58, 59n, 62, 66, 111, 116, 136
Enggass, P. M., 89n
England, 48–49, 65, 104
crop patterns of, 85
"enterprise-combination" regions of, 118
service areas in southern, 120, 128
Englishmen, 75
Environment. *See also* Relationships
influence of, 7
man and relationship to, 7, 9
natural, 9
societal, 9
Environmental forces, 128
Environmentalist view, 7
"Environmentalists," 11
Enyedi, G., 27–28, 85n, 102, 109, 124
Erickson, F. C., 86
Erosion, 72, 80
man and nature in, 143
man's role in, 142
soil, 51, 55, 73, 140–41
wind, 140
Ethnic:
groups, 73, 98, 150
patterns, 72
superiority, 75
theory, 98
Europe, 20, 32, 45, 77, 83, 87, 97, 106, 148
Central, 56, 115
"agricultural zones" of, 117
farm villages and isolated farmsteads of Western and, 98
"farming system" regions of Western and, 118
smaller farms of settlers from, 73
strip lynchets in Western and, 53

Europe (*cont.*)
Eastern, 28, 92
field borders in, 78
food production areas in, 153
intensively farmed regions of, 90
irregular block field form in, 99
mountain farming in, 44
Northwestern, 39, 56, 62, 67, 82, 101
agricultural intensity in, 65–66
cities in, 60
coastal marshes of, 87–88
land use intensity in, 63
net gain in, 63
ridges and furrows in, 49–50
spread of terracing into, 51
tenacity of the village in, 105
plains in, 34
political network of, 85
ridged fields in, 47
scholars of, 123
Western, 55, 115
farm villages and isolated farmsteads of Central and, 98
"farming system" regions of Central and, 118
regional specialization in, 64–65
strip lynchets in Central and, 53
European(s), 18, 154
colonialism, 162
expansion, 147
farmers, 72
flood tide of, 97
western, diet, 9
Europeanization, 97
Evans, E. E., 47, 48n
Evaporation, 25
Evapotranspiration, 33, 38
Everglades, 37
Evers, M. W., 105
Exploitation:
colonial, 147
destructive, 147
sequence of, 142
Exports, 119
Extensive agriculture, 69
Eyre, J. D., 119

FACTORIES, 44
Fallow period, 39
Familial lineage, 81
Families, 154
Farber, S. M., 156n
Farm(s), 24–25, 66, 106, 128, 132
abandonment of, 44
animals, 157
beginnings in South Australia, 101
calendar, 76
California, 106
citrus, 37
clusters, 132
communistic, 83
consolidation of smaller, 66, 87–88, 104
cooperative, 61, 83
cotton, 37, 86
crop, 66
dairy, 66, 102, 107
data, 116
English, 38
family, 104
fragmentation, 78, 90
fragmented, 104
French, 48
German, 38, 48
individual types, 81
inefficiencies of small, 61, 88
isolation of, 134 (fig.)
laborers, 40
large, in subtropical southeastern United States, 102
larger and more compact, 103
livestock, 38
machinery, 49, 66, 146
migration, 148
operational structure of the, 120
operators, 144, 149
population, 22–23, 28, 43
production, 146
products, 153
radial forest, 61
records, 118
settlements, 74, 103
size, 28, 51, 79, 88, 115, 128, 153
smaller, of settlers from Central Europe, 73
social stratification in settlements associated with large, 81
Soviet, 81, 83
structure, 116
tenant-operated, 80
tobacco, 86, 112
tools, 104
type, 122
vegetable, 37, 128
villages, 77, 80–81, 98, 105, 108
zones of intensity around, 61
Farmer(s). *See also* Agriculturists; Peasants; Planters; Renters; Rural man
absentee, 80
American, 76
in the American South, 72
of Anglo-Saxon origin, 73
Appalachian, 82
attitudes toward droughts, 149
attitudes toward frontier farming, 82
Belgian, 50
Canadian, 132
commercial, 82
in the Corn Belt, 64
cotton, 86
dairy, in southern California, 61
in the driest areas, 149
Dutch, 76
English, 75
European, 72
Finnish, 75, 132
French, 50
German, 49, 73, 75
Ghanaian, 70
grain, 95, 149
immigrant, 72, 74, 76
indigenous, 83
irrigation, 142
Israeli, 81
of Japanese extraction, 74
large, 80, 145
large full-time, 80
livestock, 142
local, 136
located immediate to water bodies, 47
in lower Alsace, 49
migration of, into the cities, 147–48
Minnesota, 140
native, 76
in the Northeastern United States, 44
Oriental, 39
part-time, 80
Polish, 74
psychological aspects of, 82
reactions of, to differentials in return 109
resident wheat, 80
rice, 77
Scandinavian, 74
Scotch-Irish, 75
share, 79
small, 82
small full-time, 80
smaller, 79–80
southern California, 150
subsistence, 82
successes and failures of, 130
"suitcase," 80
tenant, 79
urban fringe, 67
welfare of, 15
wheat, 80
Farmhouses, 19, 27
Farming, 64, 100, 145
Amish, 77n
belief in superiority of German, 75
boundary, 131
calendar, 38–39
communities, 144
complex mixed, 119
contrasts in, fortunes, 75
credit, 72
crop, 123
cycle(s), 37, 76
dairy, 109, 123
decline in, activity, 145
disruption of, activities, 144
dry, 102–3
field-crop, 59–60
frontier, 82
general, 60
grain, 80, 95, 123
grain-cattle, 60
grain-hog, 60
of hilly areas, 44
implements, 162
inertia of, 103
land use influences prejudicial to, 144
landlord-tenant, 72
large-scale, 80
livestock, 41, 80
mechanized, 104
methods, 9, 97, 162
mountain, 44
operations, 15, 37, 76
outliers, 134
pasture-livestock, 60
pioneer, in the Peace River Valley, 73
polar, fringe, 132
policies, 85
practices, 155
in Quebec, 39
rationalization, 102
specialization, 44
speculative, 74
system(s), 18, 51, 61, 74, 103, 114, 116

ning (cont.)
stem(s) (cont.)
cash-grain, 64
corn-hog-beef cattle, 64
differences in, 85
distortion and obscuring of, 118
distributional patterns, 56
English, 21
maps of, 24
as an occupation rather than a way of life, 82
regionalization of American, 117–18
regions, 113, 115–19
spread of, 98
total, 115
traditional lag of, 103
as a way of life, 72, 150
in the Yamato Basin, 35
chnology, 37
adition, 82
pes, 20, 111, 114, 136
alue added by, 22
ear:
monthly progression of labor demands during the, 25
operational complexities of the, 38
regional comparisons of the, 38
nland:
bandonment of, 103
iminishing, 144
ss of, to expanding cities, 143
eservation of all good, 146
nstead(s), 6, 41, 106. See also Barns; Buildings
bandonment of, 106
uildings and time-and-motion studies, 67
ombinations with agricultural systems, 20
omparison of, 20
airy, 67
istance between field and, 66
istribution of, 132
ccentric position of the, 66
rowth rates of, 108
olated, 98
aps of, 24
ersistence of smaller, 105
pie," 61
ampling of, 19
tudies, 66
ansportation-oriented, 67
cher, D., 6–7, 103, 148, 149n
cett, C. B., 152
vre, L., 8
d, 106
ahin, 77
, E., 45, 54, 131n, 133n, 140n, 148, 149n, 156
ces, 19
élon, M. P., 53n
ton, B. L., 27n, 115n
ility, 47, 55, 140, 155
ilization, 95
ilizer(s), 148, 162
eler, P., 78
d, N. C., 128
d(s), 115. See also Plots
ncient, 20
"arched," 48
orders, 78
carting manure to the, 61
consolidation of, 105
contrast between open and enclosed, 78
distance between farmstead and, 66
extensive use of machinery in large, 77
form(s), 18, 97, 104–5, 112
block, 99
irregular block, 99
original centers and dispersal routes of, 98
-regions, 112
strip, 99
fragmented, 61
idle, 108
large and squarish, 104
of marked irregularity, 107
monthly distribution of farming activities over the, 38
observations, 28
outlines, 24
patterns, 34, 90
imprint of religion on, 77–78
intersection of drumlins and, 140
ridged, 47–48, 53
rotation of uses over the, 114
sectorial, 61
settlement-, relationships, 108
shifting of, 115
smaller, 78
system regions, 113–14
systems, 114–15
three-, rotation, 58 (fig.)
wet, terracing, 52 (fig.)
work:
aerial photographs, 19–20
and archival research, 20–23
cooperative, among Eastern European countries, 92
mapping of small areas, 17–18
methodological justification, 17
questionnaires, 19
sampling, 18
Fielding, G. J., 87
Fifer, J. V., 132
Financing policy, 87
Finch, V. C., 14, 18, 27
Finkel, H. J., 80
Finland, 33, 132
Finley, R., 18, 19n
Fischer, A., 152, 154
Fishing, 123
Fite, G. C., 101, 102n
Flax, 60
Fleming, L. F., 77n
Flint Hills-Bluestem pasture lands, 79
Flocks, 41, 142
Floriculture, 60. See also Horticulture
Florida, 40, 140
Flowering, 32
Flood(s):
dangers of, 146, 149
lands with high frequency of, 149
protection, 149
plains:
agricultural occupance of, 149
classification of, 36
in Japan, 34
in the United States, 149
runoff, 143
Fochler-Hauke, G., 93n, 108n
Fodder crops, 102, 115, 117
Fonaroff, L. S., 161, 162n
Food(s), 148
capacity of the earth to produce, 151
commercialization of basic, 162
crop production, 157
crops, 153
deficiencies, 155
economy of each country, 161
improvement of supply, 162
less nutritive, 162
"non-protective," 160–61
preferences, 162
-processing industries, 69
production, 156, 162
production areas in Europe, 153
"protective," 161
requirements of the individual, 153
sufficiency, 160
supplied a single market, 119
tastes in, 162
world's, problem, 156
Food and Agricultural Organization, 119, 160
"Foodcrop climates," 32
"Foot plow," 100
Forage, 36
Forelands, 135 (fig.)
Foremen, 81
"Foreposts":
agricultural, 132
settlements lying beyond the agricultural, 134
Forest, 87, 120
-agricultural boundary, 132
boundary, 135 (fig.)
boundary between cropland and, 131
Canadian and Finnish boreal, lands, 128
intrusions into the, 133
readvance of the, 106
Forestry, 58 (fig.)
"Form elements," 10
Forms, 121
"combination," 124
"economic," 123–24
"functioning," 120, 123
settlement, 97–98
Forward, C. N., 20, 21n
Fossils, 48
Fractional code, 18, 27, 124
Fragmentation:
of agricultural land, 144
farm, 78, 90
"rational," 104
France, 29, 102, 105
Francis, K. E., 132
Freight rates, 63
Frémont, A., 80
French, 91
Canada, 99
living standards, 152
settlers, 50
Frequency distributions, 22
Fribland, V. M., 130
Friedrich, E., 3–4, 142
Fringe. See also Boundary; Frontier
agricultural, 133, 136
"highland boundary," 134
polar farming, 132
rural-urban, 150

"Fringe zones," 133
Frontier. *See also* Boundary; Fringe
  advance of the agricultural, 89
  agricultural, 134 (fig.)
  agricultural islands on the, 132
  agricultural settlement, 15
Frost(s), 45–46
Fruit(s), 39, 59–60, 79, 101–2
  Canada, major area of, 144
  patterns of ripening, 40
  regions, 115
  "tender," soils, 144
Fuels, 154
"Full ecumene," 133
Functions, 118
  rural, 120
  supply and demand, 119
  that give characteristics to a region, 125
  urban, 120
"Functioning forms," 120, 132
Fundamentalist sects, 77
Furrows, 49

GAFFNEY, M., 145–46, 147n
Gagnon, J., 136
Gallatin-Absaroka Range, 135 (fig.)
Game Theory, 70
Garden culture, 123
Garnett, A., 36
Garrison, W. L., 29n, 70n
Gasson, R. M., 67
Gathering, 100
"Gatherings," 48
Gaul, 1
"General nutrient balance," 161
Gentilcore, R. L., 101
Geographic botany, 3
"Geographic economic information," 4
"Geographical political economy," 4
Géographie active, 16
Géographie appliqué, 16
Geography. *See also* Agricultural, geography; Economic, geography
  agrarian, 6–7
  human, 31
  macroeconomic, 13
  microeconomic, 13
  "pure," 16
  "qualitative," 6
  "quantitative," 6
  relationships with economics, 3–4
  scientific, 11
  as service to mankind, 16–17
  settlement, 6
  social, 6
  theoretical, 13, 26
"Geography of dietary regimes," 162
"Geography of nutrition," 160
"Geonomics," 7
George, P., 9–10, 25, 79n, 81, 124, 148, 149n
Georgenberg, 79
Georgia, 104
Gerasimov, I. P., 141n
Gerling, W., 102
Germania, 1
Germanic:
  areas, 98
  peoples, 98
German(s), 72, 151
  agricultural regions, 129
  belief in farming superiority of, 75
  failure in southern Brazil of, 74
  farmers, 49, 73, 75
  farms, 38, 48
  immigrants, 74
Germany, 29, 48, 57, 81, 91, 114, 124
  agricultural imports by West, 66
  climates of ancient, 45
  corn areas in West, 32
  crop patterns of, 85
  farmers from, 73
  farming in West, 66
  "historical-social" agricultural regions in, 24
  lowland northwestern, 50
  North Sea coasts of, 54
  relationship of, to its neighbors, 85
  rural society of, 79
  "social fallow" in West, 106
  villages in northwestern, 105
  West, 32, 153
Ghana, 61
Gibson, J. S., 122
Gillies, J., 146
Ginsburg, N., 157
Glacial landforms, 35
Glaciation, 34
Glauert, G., 108
Glendinning, R. M., 142, 153
Goats, 95
Goldthwait, J. W., 106
Gosling, L. A. P., 79, 157
Gottmann, J., 69n, 119
Götz, W., 7, 14
Gould, P. R., 70
Gourou, P., 55, 155
Government programs, 86
Governmental controls, 84
Grain(s), 39, 60, 100–2, 106, 156
  area, 21
  -cattle farming, 60
  crops, 111
  cultivation, 95
  culture, 95
  extensive reductions in acreages of, 86
  farmers, 95, 149
  farming, 80, 95, 123
  -hog farming, 60
"Grain equivalents," 22
Grams, 39
Granaries, 154
Grape-packing houses, 106
Graphic Summary of the U.S. Census of Agriculture, 26
Graphs:
  bar, 25
  "climo-," 25
  line, 25
  radial, 25
  scatter, 25
  showing seasonal progression, 38
  two-dimensional, 25
Grasslands, 45
Graziers:
  seasonal movements of, 40
  in the southern part of Bahia, 87
Grazing, 58 (fig.)
Grazing areas, 66
"Grazing economies," 115
"Grazing pressure," 143
Great Lakes cutover areas, 139
"Great Lakes-to-Gulf Profile," 18–19
Great Plains, 32, 40, 45, 88, 104
  Central, 86
  drought hazard in the, 149
  resident wheat farmer in the, 80
  wind erosion of the, 139
  winter wheat region of the central, 1
Greenhouses, 60
Gregor, H. F., 9, 19, 22n, 26n, 40n, 4 56, 61n, 62n, 69n, 74, 80, 8 87n, 102, 106, 109n, 119, 12 124, 127 (fig.), 128, 132n, 142, 1 145n, 146, 147n
Griffin, P. F., 145
Grigg, D., 123n, 125
Gross:
  income, 116–17
  value, 119
Grotewold, A., 59, 65, 66n, 67, 85
Growing season, 39
  length of, 115
  long, 154
Guadalquivir Delta, 88
Guadeloupe, 100
Guérémy, P., 79
Gulick, L. H., 88n
Gullying, 143
Gvozdetskiy, N. A.,114n

HABITAT, 9
Haciendas, 105
Hagerstrand, T., 70
Hahn, E., 95, 99–100, 123–24
Hale, G. A., 20–51, 52 (fig.)
Hall, P., 59
Hall, R. B., 35, 81
Hambloch, H., 131, 133–35
Hamburg, 59
Hamlets, 81
Hammond, E. H., 124, 126 (fig.)
Hance, W. A., 112n
Hanson, E. P., 155
Harms, B., 4
Harris, C. D., 55, 141n, 156
Harris, D. G., 95
Harris, R. C., 90, 146
Harroy, J. P., 147
Hart, J. F., 19, 25, 43, 77, 78n, 82, 86, 87 90, 104n, 106, 107n, 112, 146
Hartke, W., 50, 101, 102n, 105
Hartshorne, R., 12, 13n, 22–23, 25, 1 121n, 125, 136
Harvesting, 39
Harvey, D. W., 13, 29n
Hatt, G., 99, 100n
Haudricourt, A. G., 49, 78
Hay, 60
Haystead, L., 101, 102n
Health Organization of the League of N tions, 160
Hearth:
  agricultural, 94, 99
  Southeast Asian, 95
  Southwest Asian, 95
Hedgerows, 78
Hédin, L., 78
Helbaek, H., 95
Henderson, J. H., 109

equen, 105
ding, 95, 100, 123
ds, 41, 142
ner, A., 3–4, 8, 10
ves, L., 86
ore, J. J., 34, 36
rarchization, 125
rarchy(ies):
f regions, 119, 125
ocial, 79
patial, 129
rsemenzel, S.-E., 25, 38
bee, E., 69, 87, 148
h Plains, 109
ghland boundary" fringe, 134
hsmith, R. M., Jr., 24n, 130
and mountain lands, 94
man, R., 2, 15
ed labor, 66
torical:
approaches, 99
determinism, 142
development, 18
perspective, 94
elationships, 9
tory, 97
ag, L. P., 27n, 115n
e, 51
e culture, 99–100, 123
fman, L. A., 119
fmeister, B., 41–42, 87
kkaido, 90
lstein, W., 113, 152, 156, 157n, 158 (fig.), 160
mestead, 41. See also Farmstead
re, P. N., 31, 32n
rnberger, T., 79–80
rses, 60
rticultural:
production, 69
products, 66
rticulture, 34, 59–60, 123
rvath, R. J., 63, 64n, 129
uses, 66. See also Buildings; Farmsteads
brooder, 106
chicken, 106
combined, and barns, 107
distribution of, 81
farm, 19, 27
grape-packing, 106
quality and style of, 81
tenant, 106
y, D. R., 100, 101n
dson, G. D., 18
man forces, 120, 128
man geography, 31
mboldt, A. von, 1
midity, 45
ngarians, 72
nter, J. M., 61n
nters, 100
nting, 100, 123
nting-collecting, 100
ntington, E., 33, 34n, 142, 148, 160–61
rwitz, N., 59, 60n
tterites, 77
draulic installations, 88
droelectric installations, 135 (fig.)

ELAND, 31

Ignorance, 162
Ilesic, M. S., 98
Illinois, 105, 122
Immigrant(s):
farm groups, 72, 74–77
German, 74
Immigration policies, 156
Imperial Valley, 40, 91
Imports, 66, 85, 119
Income(s), 66
levels, 162
per capita, 157
India, 51, 81
Indian:
crop distributions, 21
living standards, 152
Ocean, 51
-Pakistani partition, 85
Indiana, 101
Indochina, 51
Indochinese farming, 51
Indonesia, 51, 83
Industrial:
crops, 102
development, 154
establishments in villages, 148
plants, 136
revolution, 153
worker, 82
"Industrial finishing," 60
Industrialization, 134, 154
Industry(ies), 148
foreign cheese, 73
large populations and, 154
livestock, 8
promotion of, 146
raisin, 121
Inertia, 103
Inheritance, 78
Innovations, 70, 109
Insolation, 50
Instrumentation, 47
Intensification, 55
Intensity, 117, 145
capital, 63
gradations in farming, 134 (fig.)
high agricultural, 65
labor, 63
land use, 63
of land use, 9
of land use and production, 120
low farming, 133
value, 116
worldwide pattern of, 63
zonation of, 67
zones in production, of farming, 136
Intensive subsistence tillage, 123
Intercropping, 38
International Geographical Union, 157
"Invention," periods of, 103
Inventories, 157
Inventorying, 157
Iranian-Mesopotamian area, 94
Ireland, 47, 87
Irrigable lands, 130
Irrigated land, 130
Irrigation, 85, 104, 141
acreage available for, 156
actual limits of, 130
adoption of, 109
advances of, 103
canals, 54
farmers, 142
and ridging fields, 47
in the San Joaquin Valley of California, 80
Isaac, E., 78n
Islamic calendar, 76
Islands, agricultural, 132–33, 135
Isle of Man, 19
Isochrones, 60–61
"Isolated state," 57, 59, 61–62
"Isolated world state," 62
Isolation, 36, 133
Isotherm, 130
Israel, 61
Italians, 72–73
Italy, 50, 88
Ivanova, E. N., 130
Iwaki Basin, 35
Iwata, M., 73

JACKSON, W. A. D., 130
Jager, F., 133
Jager, H., 48
Jamaica, 150
James, P. E., 13n, 73–74, 91, 155
Japan, 34–35, 51, 56, 81, 153–54
decreasing immigration from, 76
irrigation canals in, 54
Japanese, 155
in California agriculture, 73
colonies in Brazil, 76
dietary demands of, 151
farmers of, extraction, 74
living standards, 152
prefectures, 119
Jefferson, M., 63, 64n
Jenks, G. F., 26, 66, 80
Jensch, G., 25, 38
Jensen, J. G., 147n
Joerg, W. L. G., 64n, 73n, 131n
Johnson, H. B., 58 (fig.), 59, 74, 140
Johnson, J. R., 29n
Jonasson, O., 59–60, 123n
Jones, C. F., 123n
Jones, W. D., 17–18, 22, 24n
Journaux, A., 87n
Juillard, E., 49, 103n
Jura, 35
Jute, 87

KALAMAZOO, 108
Kansas, 79
Kaups, M., 105
Keasbey, L. M., 4
Kelantan Delta, 34, 122
Kellogg, C. E., 56, 147
Kentucky, 106, 122
Kerr, D., 35
Kerridge, E., 49
Kittler, G. A., 48–49, 51, 53, 54 (fig.)
Kniffen, F., 99
Kohn, C. F., 20
Kollmorgen, W. M., 66, 72–75, 76n, 77n, 79–80, 149
Korea, 51
Kostrowicki, J., 92, 157
Kotschar, V., 112n
Kovalev, S. A., 92

Kramer, F. L., 100n
Krenzlin, A., 104n, 105, 136
Krueger, R. R., 35, 115, 144
Kryuchkov, V. G., 92, 114n
Krzymowski, R., 2, 14–15, 59
Küchler, A. W., 142
Kühn, F. H., 60
Künzler-Behncke, R., 91

LABOR:
human, 82
indices, 118
intensity, 63
productivity of, 109
shortage of, 149
surplus, 80
"Labor hour," 118
Laborers, 81
Lacustrine plains, 35
Lambert, A. M., 88n
Land(s), 150
abandonment, 146
agricultural, 116–17, 144
arable, 131, 149, 151–52
assessment, 114
attitudes toward money and, 82
best, 156
better-drained, 132
capability regions, 113–14
in cash agriculture, 157
cheap, 144
competition for cheap, 144–46
contraction of agricultural, 44
costs, 104
cultivated, 133
derelict rural, 139
desert, 151
developing new, 156n
draining of, 45
dry, 88, 103, 130, 133
evaluation of the, 151
exorbitant rates of, 162
expansion of agricultural, 152
extent of good, 155
fragmentation of agricultural, 144
high-price, 144
higher prices of, 146
highly prone to erosion, 141
hill and mountain, 94
increase in improved, 85
inheritance system, 18, 48
intensive inventories of, 157
irrigable, 130
irrigated, 130
leasing of, 105
loess, 94
loss of agricultural, 146
man, relationships, 147
mountain, 36, 79, 133
need for, 83
negligent use of the, 139
as an objective in itself, 82
overcrowded, 156
ownership status, 82
pasture, 79
periodic redistribution of, 98
potentially arable, 131
pressure on the, 147
productive capabilities of agricultural, 153
productivity of the, 109
reclamation, 88, 141
reforms, 83
registry records, 79
as a resource of man, 147
resource regions, 114
roots, 82
sinking of surfaces of, 146
with soils conserved or improved, 142
spatial zonation of agricultural, 62
sprawl of cities over agricultural, 145
steppe, 151
supply of, 42
surveys, 89
tax, 80
tenure, 9
terraced, 34
transfer, 144
urban encroachment on agricultural, 144
use(s), 43, 114, 128
amount of area affected by changes in, 46
Canadian, survey sheets, 157
closer adjustment to the environment, 44
declining intensity of, 106
on glaciated plains, 35
impact of survey systems on current, 89
influences prejudicial to farming, 144
jointly optimum intensity of, 29
of large and small landowners, 79
map(s), 21, 23, 79
mapping, 92
new, introduced into abandoned areas, 86
patterns, 13, 19, 35, 42, 72
"periods," 100
on the Prairie du Chien terrace, 34
sequence of gradation in, 63
shift to ever more intensive, 106
spatial and chronological progressions on a farm, 25
surveys, 26
systems, 115
trends, 109
vertical zones, 131
World Land Survey, 26, 157
zonation near cities, 67
zonation on alluvial fans, 35
use of the best, 162
value of the, 149
variations in the quality of, 151
in want, 142
wet, 88
"Land economy," 122
"Land types," 114
Land Utilization in East-Central Europe: Case Studies, 92
Landforms, 28, 47
Landholders, 79
Landholding systems, 51
Landholdings, 80
Landlords, 81–82
"lands," 49
Landscape(s):
agricultural, 65, 78, 94, 107, 139, 145
agriculture which is expressed in the, 157
appearance, 121
changes in an agricultural, 109
coconut, 129
"constituting components" of the agric tural, 120
crop, 38
cultural, 6
economic, 10
effects of political decisions on the rur 84, 87–88, 92
elements of the agricultural, 112
English agricultural, 85
Europeanization of the, 97
hoe culture, 99
"polder," 122
a receiving, 97
reflection of taboos in the agricultur 57
regional expressions in function and, 1
regionalization of crops and animals the agricultural, 112
rubber, 129
rural, 10, 76, 84, 92, 103
structural characteristics of the agric tural, 97
Thünen circular pattern in the, 59, 69
treatment of the, 141
Langellé, M., 161
Languedoc, 100
Large, D. C., 86n
Latin America, 42, 47, 83, 147, 160. Se *also* South America
farm types and associated settleme structures in, 81
tribal areas of, 81
vitamin deficiencies in, 161
Laur, E., 60, 66
Law(s), 3, 11–13
of diminishing returns, 55
Piedmontese, 88
state marketing, 87
Leasing, 105
Lee, C., 35
Lee, D. H. K., 155
Legumes, 36, 39
Leighton, P. A., 147n
Lemon, J. T., 75
Leopold, L. B., 140
Leser, P., 100
Lessinger, J., 146
Levees, 34
Lewis, R. A., 130
Lewis, W. A., 154
Lewthwaite, G. R., 65n, 74, 128
Liesegang, F., 32
Life-curve, 140
Lindberg, J. B., 15, 28n, 125–26
Lipschütz, A., 160n
Lisbon, 157
Livestock, 101, 117, 143. *See also* Animals Cattle; Sheep
distribution of, 36
farmers, 142
farming, 41, 80
industries, 8
raising, 9, 62, 86, 95, 102, 123
seasonal movements of, 40
significance of the mountain-valley com plex to, 35
systems, 65
Living standards, 37, 146
French, 152

ng standards (*cont.*)
ndian, 152
apanese, 152
preserving, 156
variation in, 151
ss lands, 94
an, R. F., 135n
don, 59, 67
g lot(s), 90–91, 99
Angeles, 62n, 119
County, 69
milkshed, 87
ering, J. H., 63, 64n
venthal, D., 46n
ver California, 143
bke, B. H., 82
gens, R., 8, 10, 123n

AODKA, B. S., 87
Carty, H. H., 13, 15, 124, 125n, 136
Cune, S., 100
Dermott, G. L., 89
cGregor, D. R., 36
Kain, W. C., Jr., 83
ckintosh, W. A., 63, 64n, 73, 131n
cNeish, R. S., 97
chinery, 44, 148
extensive use of, in large fields, 77
farm, 49, 66, 146
sheltering of, 106
strictures on the use of, 77
chines, 102
croeconomic geography, 13
dagascar, 51
dura, 55
itland Flats, 122
ajor agricultural regions of the earth," 123. *See also* Areas; Regions
laya, 34, 40, 157
laysia, 79
lin, J. C., 149
linowski, B., 76, 77n
lmström, V. H., 31, 32n
lnourishment, 160–62
lnutrition, 160
lthus, T. R., 151
althusianists," 163
n-environment relationships, 45. *See also* Relationships
n-land relationships, 147. *See also* Relationships
n-made lakes, 54. *See also* Reservoirs
an-day(s)," 22, 118
an-work hours," 38
ngo groves, 144
nitoba, 89, 105
nley, V. P., 22n
nure(s), 61, 162
p(s), 27, 81, 102, 105, 124, 130, 156, 161
Bowman's, of "pioneer belts," 133
British Ordinance Survey, 19
"chorographic-compage," 25
choropleth, 24
diagrammatic, 24
distribution, of major crops and animals, 111
dot, 23–24
European agricultural, 24
of farming systems, 24
isopleth, 23–24
land use, 21, 23, 79
mathematics and, 29
of nutritional adequacy, 161
phenological, 32
of the polar settlement fringes, 133
of rural morphology, 24
special-purpose, 23
in strip form, 23
with superimposed diagrams, 26
thematic, 23
world land use, 157
worldwide agricultural, 157
Mapping:
of agricultural islands, 132
characterization of an agricultural region, 18
of crop landscapes, 38
fractional code system of, 18
international, program, 157
land use, in Poland, 92
technology, 26
time needed for, 17
uniform, 124
Marble, D. F., 29n, 70n
Marbut, C. F., 55
Market(s), 37–39, 44, 57, 62, 64–65, 121
gardening, 58 (fig.)
for the major agricultural commodities, 119
rotation systems near centers, 115
Marschner, F. J., 36, 90, 112
Marsh, G. P., 45, 46n, 47, 140n, 147
Marshes, 88
Martonne, E. de, 8n
Marxists, 141, 163
Massachusetts, 35
Mathematical:
methods, 26, 31
models, 28–29, 69
terms, 11
Mathematics, 26
descriptive, 27
and the map, 29
theory-building, 27
Mather, E. C., 19, 25, 77, 78n, 105–6, 108, 112
Matley, I. M., 42n
Maunder, W. J., 31, 32n
Maxwell, J. W., 37, 133n
May, J. M., 161–62
Mead, W. R., 33, 34n
Mean values, 23
Meat, 59
Mechanization, 107
Mediterranean:
agriculture, 51, 123
area, 101
Basin, 50–51
Mehren, G. L., 156n
Meinig, D. W., 90, 91n, 100–101, 128
Meitzen, A., 98
Mennonites, 77
Mennonite settlements, 105
Mental activity, 155
Merner, P. G., 40, 41n
Metabolic levels, 161
Meteorological observations, 47
Metzler, W. H., 40n
Mexican Government, 88
Mexicans, 88
Mexico, 50, 97, 132
Michel, A. A., 85n
Michigan, 108
farm properties in northern, 105
swampy soil in, 75
Microclimates, 46
Microclimatic controls, 38
Microeconomic geography, 13
Middle Ages, 61
Middle Atlantic States, 104
Middle East, 50–51, 95
Middle latitudes, 97, 102, 152
Midlands, English, 79
Midwest, 22, 27, 39–40, 63, 104, 115
Migrants, 40, 148
Migration(s):
Christian, 78
human, 95
Jewish, 78
out-, 148
rural-urban, 148
Mikesell, M. W., 78n
Military needs, 147
Milk, 60
market, -cheese-butter zonation, 65
production in northern Ireland, 87
production in the Northeastern United States, 64
production in the vicinities of cities, 64
Milking parlor, 67
Milksheds, 64, 119, 136
Millington, B. R., 34, 35n
Mineral discoveries, 136
Minerals, 36, 160
Mining:
installations, 135 (fig.)
towns, 136
Minnesota, 140
Mississippi:
River, 43
share and tenant farmers in the lower, Valley, 79
spread of soybeans into the lower Valley, 86
Valley, 39
Missouri, 73
Mittelbach, F., 146
Models, 13
behavioral, 70–71
building and testing, 69
dealing with behavioral problems, 29
dealing with economic problems, 29
mathematical, 28–29, 69
multiple regression, 31–32
probability, 109
ring-type, 67
spatial, 71
static equilibrium, 29
stochastic, 29
theoretical, 109
Thünen, 70
Moisture, 42, 45, 56, 128, 132, 152, 154
Monbeig, P., 87
Monetary:
information, 22
values, 116
Money, 82
Monheim, F., 132
Monoculture, 27, 80, 119
Montana, 135 (fig.)

Mountain:
  forelands, 135 (fig.)
  lands, 36, 79, 133
  -valley complexes, 35
Mukomel, I. F., 114n
Mulberry, 97
Müller-Wille, W., 60, 131, 153
Multiple:
  cropping, 38, 44
  -feature regions, 113, 118
  -purpose plant, 65
  regression models, 31–32
"Multiple nutrient production relative," 161
Murdock, G. P., 112n

NAPOLEONIC WARS, 101
Narr, K. J., 95
National:
  character, 75
  policies, 85
Nationality, 72, 82
"Natural-economic zones," 114
"Natural" features, 17
Natural forces, 120
"Natural land divisions," 114
Natural regions, 114
Naylon, J., 88n
Net agricultural values, 145
Netherlands, 54, 76, 91
Nevada, 134
New England, 40
New Hampshire, 106
New Jersey, 128
New South Wales, 91, 122
New World, 95, 97
New York, 106
New Zealand, 31, 97, 128
Newcomb, R. B., 20
Niagara Fruit Belt, 144
Niemeier, G., 91, 99
Nikishov, M. I., 26n, 92n
Nitrogen, 36
Noh, T., 54
Nomadic herding, 123
Nomadism, 41, 77
Nomads, 41, 115
Norden, 134 (fig.)
Normandy, 80
North America, 2, 18–19, 45, 49, 67, 83
  arable land resources of the Soviet Union and, 128
  long-lot system in, 90
  polar agricultural margin in, 135
  polar settlement fringes in, 133
  rectangular system in, 90
  scholars of, 123
North Sea coasts, 54
Northwestern University, Geography Department, 18
Norway, 47
Notestein, F. W., 153, 154n, 162
Nova Scotia, 27
"Numerator elements," 18
Nurseries, 59
"Nutrient production relative," 161
Nutrient(s), 47, 161
Nutritional:
  adequacy, 161
  levels, 162
  quality, 162
  requirements, 162
Nuttonson, M. Y., 32, 113

OASES, 22, 132
  labor force in the, 146
  of the southeastern deserts, 102
  Western, 63
Oats, 60
Obst, E., 60n, 62, 124
"Occupance types," 36
Ohio, 90, 98
Old World, 95, 131
  agricultural areas of the, 107
  attitudes toward animal foods in the, 78
  diets, 162
  longer period of stocking in the, 143
  oldest remnants of plants found in the, 97
  spread of the vine over the, 78
Olmstead, C. W., 22n, 115
Olson, C. E., Jr., 20n
Onondaga district, 106
Ontario, 132, 140
  Clay Belt, 89
  Dutch farmers in southwestern, 76
  Government, 89
  Niagara Fruit Belt of, 144
  Shield, 37
  swampy soil in Michigan and, 75
  type-of-farming areas, 19
Opportunity costs, 150
Orange groves, 144
Oranges, 40
Orchards, 35
Organic materials, 95
Osborn, E. P., 141
Osiers, 59
Otremba, E., 2, 6–7, 9–10, 18n, 24–25, 26n, 33, 34n, 61n, 62, 80n, 81, 83n, 85n, 97, 104, 107, 112, 115, 119, 123, 124n, 125, 128n, 129, 148, 149n
Outliers. *See also* "Foreposts"; Fringe
  agricultural, 133
  farming, 134
Overgrazing, 142

PACIFIC, 51, 95
Pacific Northwest, 40
Pacific Southwest, 40
Pakistan:
  cotton in West, 87
  jute in East, 87
  relations between the two parts of, 87
  rice farmers of East, 77
Palm kernels, 84
Pampa, 56, 90
Papadakis, J., 32, 113
Parallel invention, 97
Paris, 69, 122
Paris Basin:
  leasing of lands in the, 105
  rural social types and property sizes of the, 79
Parsons, J., 47, 48n
Passarge, S., 7
Pasture(s), 115, 117, 120
  Flint Hills-Bluestem, lands, 79
  intermediate, 41
  -livestock farming, 60
  mountain, 41
  seasonal, 41
  summer, 41
  summer mountain, economy, 40
Patagonia, 60
Pattison, W. D., 90
Patton, D. J., 24n
Peace River:
  area, 63
  pioneer farming in the, Valley, 73
  pioneer fringe, 19
  settlement fringe in the area, 133
  townships in the area, 132
Peanuts, 84
Peasants, 82, 148. *See also* Agriculturists; Farmers; Renters; Rural man
  German, 75
  North European, 61
Peattie, R., 35, 131
Pelzer, K. J., 40, 133
Penck, A., 152, 154
Pennsylvania, 75
Pennyroyal, 122
Peplies, R. W., 56
Perceptions, 149
Perpillou, A. V., 100
Perret, M. E., 73, 76
Peru, 89
Peterec, R. J., 112n
Pfeifer, G., 53n, 67, 69n, 79, 112, 124n
Philippines, 40, 51, 104
Phlipponneau, M., 69, 122
Photographs, 81, 133
"Photosynthesis, potential," 157
Physical:
  barriers, 9
  controls, 131
  laws, 12
  sciences, 45
Physiognomic agricultural geography, 6
Piedmontese law, 88
Pig, 78
"Pioneer belts," 133
Plains, 34–36, 54, 149
Planning, 149
Plant(s), 78. *See also* Crops
  damage by air pollution, 147n
  damage to, 146
  -growth process, 139
  industrial, 136
  movement of, within the tropics, 97
  oldest remnants of, 97
  processing, 105
  relationships between animals and domesticated, 12
  remains, 97
  routes followed by animals and, 95
"Plant-capability" areas, 33
Plantation(s), 80–81, 83, 102
  culture, 123
  distribution of tropical crops grown on, 84
  geographic reality of individual, 122
  studies, 122
  tenant houses on Southern, 106
  top social layer on the, 79
  workers, 40, 83
Planters, 79. *See also* Agriculturists; Rural man
Planting:
  origin and dispersal of, 96 (fig.)
  type, 94
Plateaus, 155

tt, R. S., 81
istocene beaches, 35
ssis de Grenédan, J. de, 15
ny, 1
ts, 69, 81, 104. *See also* Fields
w(s), 47–51, 53, 80
culture, 99, 123
"foot," 100
wing, 49, 61
dzolics, 55
ar boundaries, 130–31
settlement fringes, 133
older landscape," 122
Valley, 91
and, 157
es, 72
itical:
-agricultural relationships, 85
antagonisms, 85
decisions, 84
olitical determinism," 86
itics, 91
len, 143
pulation, 66
agglomerations, 154
calculation of world capacities, 151–52
capacity, 147, 151, 160
concentrations in tropical Africa, 155
control, 141
cycle theory, 154, 162–63
density, 9, 129
densities in each climatic area, 152
disaster for the West, 148
equilibrium, 154
estimates, 153
farm, 22, 28, 43
growing urban, 144
growth and food supply, 162
"life cycle" of rural, 107
limits, 151
over-, 153
and the physical and social environment, 154
pressure of, 141, 149
ranges for the United States, 152
rate of growth, 163
records, 153
rural, 23, 26–70, 40, 73, 77, 107, 121, 129, 148, 153
share of the agricultural, 119
stress, 154
under-, 153
opulation-resource ratio," 157n
rk, 78
rtuguese, 74
ossibilism," 8
ossibilist" concepts, 8
tatoes, 31, 60
verty, 162
airie du Chien terrace, 34
airie Provinces, 130
airies, Canadian, 63, 89
ecipitation, 28, 140
ediction, 27, 108–9
efectures, Japanese, 119
ice, D. A., 77n
ice-support programs, 149
ices, agricultural, 148
icing, governmental, 87
rinciples, 112, 120

Processing plants, 105
Produce, 119
Products:
disposal of the, 120
farm, 153
truck, 60
"Product Information," 3
Production:
caloric, 113, 153, 156
farm, 146
milk, 64, 87
periodic shifting of centers of, 39
Productivity patterns, 55
Programmers, 29
Protein(s), 160–61
Proudfoot, M. J., 18, 19n
Prunty, M. C., Jr., 21, 23n, 43, 79, 81n, 86, 101–2, 109, 122, 136
Psuty, N. P., 144
Psychology, 72
Puerto Rico, 18
Pulses, 39
Putnam, D. F., 140

"QUALITATIVE" GEOGRAPHY, 6
Quantification, 11
Quantitative:
analysis, 131
methods, 13
"Quantitative" geography, 6
Quebec, 136
farming in, 39
Government, 89
long-lot system, 90
Queensland, 91
Questionnaires, 19–20, 149

RACE, 72, 82
Racial:
characteristics, 36
differences, 37
Radar, Side-Looking Airborne, 20
Radiation, 33
Radiation balance, 141
Radioactive chemical element, 140
Radiocarbon, 97
Railroads, 62–63
"Railway Belt," 64
Rainfall, 45, 132
excessive, 157
followed settlement, 149
heavy, 143
intense summer, 143
relationships between rice and, 31
Raisin industry, 121
Rakitnikov, A. N., 92n, 114n
Ranching, 103
Ranching areas, 79
Raper, A. F., 83
Ratio(s), 21, 116, 136, 160–61
analyzing data obtained through, 23
applying to social aspects of agriculture, 22
of change, 27
in distinguishing agricultural regions, 22
input-output, 22
people-area, 27
sheep-swine, 27
Raup, H. F., 101
Ravenstein, E. G., 151, 152n, 154

Raw materials, 154
Raymond, C. W., 21
Real estate developers, 144
Reciprocal effect(s), 4, 10
"Reciprocal" concepts, 8
Reclamation, 152
by the Dutch, 75
of former sea bottoms and coastal marshes, 54
land, 88, 141
and settlement of the Guadalquivir Delta, 88
Reconstruction, 142
Rectangular survey system, 90–91, 140
Red River Valley, 91
Reeds, L. G., 2, 19, 35, 157
Region(s), 16, 111, 123, 128, 130, 136. *See also* Agricultural, regions; Areas
administrative, 87
arid, 146
boundaries of the, 120
burley tobacco, 122
characterization of the, 121
cheese, 122
climatic, 118
commodity-, symbiosis, 121
conceptual, 87
core of a, 130
cotton, 43
crop, 2, 9
dairy, 128
"developmental classification" of, 147
"economic matter" of, 3
efficacy of the, 125
"enterprise-combination," 118
extensive view of the, 121
farming system, 113, 115–19
field-form, 112
field system, 113–14
fruit, 115
functional, 113, 118, 136
functions that give characteristics to a, 125
"great," 119
hierarchies of, 119, 125
intensively farmed, 90
land capability, 113–14
land resource, 114
multiple-feature, 113, 118
natural, 114
overall agricultural, 120
pattern of, 116
population capacity of the earth and its, 151
rural, 122, 124
culture, 83
psychological, 83
Salt Lake, 42
of similar farming types, 114
single feature, 112
soil, 55
study of a single, 121
tobacco, 122
total, 120–22
total agricultural, 121, 124, 128
of vitamin deficiencies, 161
winter wheat, 149
world-, series, 123n
Regional:
approach, 1, 5–6, 125

Regional (*cont.*)
comparison, 125, 128
of farming years, 38
fruitful, 132
complexes, 2
delimitation, 33
generalizations, 17
specialization, 137
in the United States and Canada, 64
in Western Europe, 64–65
studies, 111
in the Alps and Jura, 35
in American agricultural geography, 121
comparative, 128
of North American agriculture, 117
single-feature, 112
system, 122
typology(ies), 113, 115, 129
"Regionalists," 11
Regionalization, 130. *See also* Classification; Zonation
agricultural, 27, 117
of agricultural potential, 113
of American agriculture, 21
of American farming systems, 117–18
of American tobacco farms, 112
attempts at, 125
based solely upon psychological characteristics, 82
comprehensive view implicit in, 122
criticism of, 125
of crops and animals, 111
of farming systems, 118
methods, 118
more localized and detailed, 83
physical-geographic, 92
process, 113
schemes of, 115, 124
schemes of rural society, 79
of supply and demand functions, 119
techniques of, 118
worldwide, 124
Regression analysis, 28–29, 32
Relationship(s):
between barley and climate, 31
cause-and-effect, 163
between cropping and livestock raising, 9
between diet and its supporting agricultural forms, 10
between domesticated plants and animals, 12
between farming operations and religious motives, 76
finding of, 157
historical, 9
man-environment, 7, 9, 44
man-land, 147
between physical environment and agricultural operation, 9, 31, 37
political-agricultural, 85
between population density and available agricultural space, 9
between population growth and food supply, 162
between potatoes and the Arctic Front, 31
between rainfall and rice, 31
settlement-field, 108
societal-morphological, 81
spatial, 10
terrain-agricultural, 36
between transportation changes and zonation of intensity of land use, 67
between United States and Canada, 85
Relief, 98, 121, 129
Religion(s), 72
imprint of, on settlement patterns, 77
influence of, on field patterns, 77–78
influence of, on village formation, 78
side effects of, 82
in the tropics, 78
Rent, 80
Renters, 80
Reservoirs, 47, 54
Resources:
agricultural, 143
misallocation of, 146
misuse of, 139
poor exploitation of, 141
and problems of underdeveloped areas, 112
soil, 141
threats and damages to agricultural, 147
Retaining walls, 51
Retrogression, 103
Rheingau, 100
Rhine:
field patterns in the lowland of, 34
hill country, 128
south-facing slopes in the, -land, 50
Rhizomes, 39
Rhodesia, Northern, 64
Rice, 31, 97, 123
Ridges and furrows, 48
Riley, M. P., 77n
Rio de Janeiro, 56
Rio Grande delta, 37
Rio Grande Valley, 120
Ripening, 32
River valleys, 94
Robinson, A. H., 12–13, 27–28, 124, 126 (fig.)
Robinson, E. V. D., 4
Robinson, K. W., 122
Rochow, E. G., 152
Rocky Mountains, 25, 135
Roepke, H. G., 136
Rome, 147
Roots, 39
Roscher, W., 59, 60n
Rose, A. J., 91
Rose, J. K., 31, 32n
Rotation:
systems near market centers, 115
systems of, 18, 115
three-field, 58 (fig.)
of uses over the fields, 114
Rousillon, 100
Routes:
dispersal, across the northern and central Pacific, 95
followed by plants and animals, 95
Roy, J. M., 88n
Rubber, 85
"Rubber landscape," 129
Rubenstein, E. S., 45, 46n
Rubin, V., 155n
Rudimental sedentary tillage, 123
Rühl, A., 7n, 12, 14, 107, 137
Runoff, 132
Rural:
economies, 148
exodus, 106
Land Classification Program, 18
landscapes, 10, 76, 84, 92, 103
man, 123
attitudes of, 149
condition of, 147
morphology, 24
mosaic, 24
population, 73, 121, 129, 148, 153
density, 26
dispersed, 27, 77
farm, 23
"life cycle" of, 107
nonfarm, 23
periodic flows of, 40
psychological regions, 83
psychology, 82
regions, 122, 124
service centers, 119–20, 128
settlement(s), 81, 85, 91–92
society, 79–80, 82
-urban fringe, 150
"Rural life," 122
"Rural man," 82
Russia, European, 116. *See also* Sov Union

SAARINEN, T. F., 149
Sacramento-San Joaquin Delta, 122
Sacramento Valley, 40
Sahara, 143
Salt Lake Region, 42
Salter, P. S., 144
Sampling, 20
dispersed-area method, 19
positioning of sample areas, 19
random, 19, 27
"stratified" version, 19
systematic, 19
traverses, 18, 20
San Diego area, 40
San Francisco Bay Area, 40, 119
San Joaquin Valley, 80, 121
Sao Paulo, 56
Sapozhnikova, S. A., 33
Sapper, K., 59, 130, 154, 155n
Sargent, F. O., 40n
Sas, A., 75n, 76
Saskatchewan, 89
"Satisficing," 70
Sauer, C. O., 17, 73, 94–95, 96 (fig.), 97 140n, 142–43, 147, 156n
Savannas, 87
Sax, E., 62
Scandinavia, 133
Scandinavians, 72, 74
Schaefer, I., 50
Schall, S. S., 32
Scharlau, K., 107n, 155n
Schlüter, O., 3–4
Schmidt, P. H., 59
Schmidt, P. W., 95, 100n
Schnelle, F., 32n
Schoenwetter, J., 143
Schott, C., 90
Schroeder, K., 37, 38n, 66, 120
Schul, N. W., 79

warz, G., 61n, 81n, 90n, 104n, 107
werz, J. N., 1
vind, M., 91
ntific geography, 11
land, 47
t, E. M., 18, 19n
t, P., 42
88
rs, P. B., 73
sonal:
imatic rhythm, 37
ifts of agricultural production, 39
asonal programming," 38
sonality, 38
imatically-determined, 39
ck of, 157
meyer, K. A., 153, 156n
ding, 39
-sufficiency, 85
ipalatinsk Oblast, 92
ple, E. S., 101
quent occupance," 100
ice centers:
istribution patterns of rural, 128
ral, 119–20
southwestern Wisconsin and southern England, 128
ement(s):
gglomerated, 98–99, 105, 108
ltitudinal boundaries of agricultural, 135 (fig.)
eyond the agricultural foreposts, 134
anadian, 91
anadian Prairie, 88–89
spersed, 99
vergence between agricultural margins and, 134
rm, 74, 103
eld relationships, 108
rms, 97–98
ture agricultural, 130
eography, 6
reater potential, 156
roup, 61
f the Guadalquivir Delta, 88
ennonite, 105
f the northern agricultural fringe, 89
atterns, 34
imprint of religion on, 77
rural, 92
rural, in the Red River Valley, 91
ermanent, 133, 135 (fig.)
olar, fringes, 133
rograms in Manitoba and Saskatchewan, 89
ecent, in Western Canada, 89
ainfall followed, 149
ecreation, 135 (fig.)
ural, 81, 85, 92
easonal, 135 (fig.)
edentary, 41
ocial stratification in, 81
emporary, 133, 135 (fig.)
endency toward the isolated, 108
ansportation, 135 (fig.)
gorous policy of, 89
hite, 152
ers, 133
ntz, H. L., 123n, 141
shko, S. I., 33
Sheep, 95. *See also* Animals; Livestock
in Britain, 87
distribution in Britain, 102
herds, 41
raising, 60
Shepherds, 41
Shifting:
cultivation, 123
cultivators, 39, 115
Siemens, G. von, 114, 128
Sierra Madre de Chiapas, 122
Sierra Nevada, 135 (fig.)
Simon, H. A., 70
Simonett, D. S., 79
Simoons, F. J., 78, 162
Simpson, E. S., 22n, 87n, 125n
Simpson, R. B., 20n
Sinclair, R., 67, 68 (fig.), 69, 145
Singapore, 40
Skibbe, B., 26n
Slavic areas, 98
Slovenia, 98
Smith, G. H., 148n
Smith, J. R., 141
Smith, T. L., 66, 90n, 148
Smolla, G., 95
Snow boundary, 135 (fig.)
Social:
amenities, 61
conditions, 104
geography, 6
groups, 80
hierarchy, 79
sciences, 14
stratification, 81
welfare, 15
"Social fallow," 106
Socialist countries, 92
Society(ies), 79–82, 148
Soil(s), 37, 42, 72, 143
alluvial, 56
availability of, 131
blowing, 80
changes in treatment of, 142
classification system, 56
compaction, 146
condition, 98, 113, 115
conservation, 74, 139
Conservation Service, 56, 114, 139
conserved or improved, 142
creep, 54 (fig.)
dark prairie, 45
deficiencies, 155
degradation, 86
depletion, 55, 142, 147
destruction, 139, 140n, 142
in the drier lands, 141
economic, 56
erosion, 51, 55, 73, 140–41
erosion and population capacity, 147
families, 114
impoverishment, 162
influence on agricultural patterns, 36
losses of the best, 144
minerals, 36
modification of, 55
moisture, 45, 56, 132
neglect of the, 143
nonvolcanic, 55
podzolic, 55–56
prairie, 56
problems of erosion, 150
quality, 36
regions, 55
removal, 142
structure, 55
swampy, in Michigan and Ontario, 75
"tender fruit," 144
texture of, 132
tropical, 55
types, 56, 114
volcanic, 55
-water capacity, 146
"Soil Classification, Seventh Approximation," 56
Soil Conservation, 139
Solar:
energy, 45
radiation, 157
Solberg, E. D., 145
Sonora, 143
Sopher, D. E., 76, 77n, 78
Sorre, M., 107, 162
South, the, 43
farmers in, 72
German settlers in, 74
immigrant groups in, 75
success of farming groups in, 80
South Africa, 82
South America, 56, 132
South Australia, 101
South Carolina, 73
South Dakota, 31
South Island, 97
Southeast, the, 39, 86
Southwest, 146
shift of cotton to the, 86
soil in the American, 143
Soviet Union, 33, 92, 114. *See also* Russia
arable land resources of North America and the, 128
black soil belt of the, 56
crop boundaries in the, 131
poleward expansion of agriculture in the, 131
Sowing type, of crop cultivation, 94
Soybean(s), 60, 86, 97, 101
Spade, 51
Spain, 88
Spate, O. H. K., 81
Spatial:
associations, 11
model, 71
relationships, 10
Spencer, J. E., 38–39, 40n, 50–51, 52 (fig.), 129
St. Lawrence Valley, 136
Stages, 101
Stall feeding, 41
Stamer, H., 59, 60n
Stamp, L. D., 19, 20n, 85, 143, 155
Starchy staples, 160
Starvation, 154
Statistical agricultural geography, 6
Statistical methods, 12
Statistics, 40
Steiner, R., 19
Stephan, L. L., 37, 38n
Stephenson, G. V., 87n
Steppe lands, 151

"Stepped statistical surfaces," 26
Stewart, J. R., 77n
Stock. *See* Animals; Cattle; Livestock; Sheep
Stockmen, 149
Stone, K. H., 133–34
Storms, 143
Strabo, 1
Strip lynchet, 53, 54 (fig.)
Structure:
  agrarian, 11, 121–22
  soil, 55
Studensky, G., 116–17
"Sub-ecumene," 133
Sublett, M. D., 65, 66n, 85
Subsidy, government, 87
Subsistence, 120
Subsistence agriculture, 82
Subsoil, 120
Subtropics, 97, 102
Sudan, 132
Sugar cane, 111
"Suitcase" farmer, 80
Summer mountain pasture economy, 40
Summer temperatures, 45
Supply and demand, 118–19, 146, 162
Supranational organizations, 85
Surveys, 18
Swamp, 132
Sweden, Middle, 29
"Swells and swales," 49
Swiss:
  dairymen in California, 76
  in the foreign cheese industry in Wisconsin, 73
Switzerland, 47
Symons, L., 15, 123, 124n
"System" of agricultural sciences, 5
System(s), 11
  farming, 18, 21, 24, 35, 51, 56, 61, 64, 74, 85, 98, 103, 114–18
  field, 113–15
  livestock, 65
  rotation, 18, 115
Systematic sciences, 5

TABOOS, 57, 78, 162
"Tapering principle," 63
Tariffs, 66, 91
Tasmania, 42
Tastes, 66
Tax(es), 67, 150
  assessments, 144
  urban, 145
Taylor, C. C., 83, 123n, 124
Taylor, G., 90, 152, 153n
Tea, 85
"Technical systems," 9
Technicians, 81
Technological capacities, 9
Technology, 12, 70, 104, 152
  agricultural, 5, 43
  farming, 37
  mapping, 26
  potentialities of agricultural, 55
Temperature, 42, 128, 152
  amelioration of, 46
  variations, 45
Temples, 77
Tenant(s), 81–82
  farmers, 79
  houses, 106
  -operated farms, 80
Tennessee, 87, 106
  burley tobacco region in, 122
  Valley Administration, 18
Terrace(s), 47, 50–51, 52 (fig.)
  "bench," 53
  diluvial, 34
  graded-channel, 53
  level, 53
  poorer soils of the, 155
Terraced land, 34
Terracing, 51, 52 (fig.), 53
  accounts of early, 53
  origin and dispersal of, 52 (fig.)
  wet field, 52 (fig.)
Terrain, 42, 113, 131
  -agricultural relationships, 36
  dependence of the transport route on, 63
  influence in agriculture, 34
  limited areas of gentle, 144
  "ridge and furrow," 48, 50
  "ridge and hollow," 34
Texas, 40, 120
Thailand, 31
Theoretical:
  economic geography, 12–13
  geography, 13, 26
  models, 109
Theory(ies), 11–12
  climatic, 143
  game, 70
  general, of location, 69
  location, 71
  Meitzen's, 98
  population cycle, 154, 162–63
  Thünen, 12, 58–59, 61–62, 65
    application to the individual farm, 60
    on continental and world scales, 60
Thoman, R. S., 8, 9n, 24n, 55–56
Thomas, D., 101
Thomas, W. L., Jr., 46n, 79n, 94n
Thompson, J. H., 38
Thompson, W. S., 153, 154n, 162
Thornthwaite, C. W., 25, 33, 46–47, 113
Three-field rotation, 58 (fig.)
Thrower, N. J. W., 90, 98
Thünen, J. H. von, 57, 59n, 62, 70–71
Thünen zones, 58 (fig.). *See also* Theory, Thünen; Zonation
  centrifugal movements of, 61–62
  centripetal movements of, 62
  coalescence of, 62
  distortion of, 62
  fragmentation of, 62
Tibet, 77
Timber, 132
Time-and-motion studies, 67
Timoshenko, V. P., 31, 32n
Tisowsky, K., 100
Tobacco, 60
  farms, 86, 112
  regions, 122
  in the Southeast, 86
Tokyo, 119
Tomato production, 40
Tools, 104, 112, 123, 148
"Topographers," agricultural, 1
Torbert, E. N., 26, 35, 37, 142, 153
Townships, 98, 132
Tractors, 66
Trade policies, 65
Tradition, 82
Traffic divides, 120
Transhumance, 40
  in Australia, 42
  complex, 41
  hibernal, 41
  inverted, 41
  inverted hibernal, 41
  in Latin America, 42
  mixed, 41
  normal, 41
  normal hibernal, 41
  partial, 41
  petite, 41
  reevaluation of the definition of, 41
  in Tasmania, 42
  tropical, 41
  in the western United States, 42
"Transhumance in the Sheep Industry the Salt Lake Region," 42
Transhumants, 41
Transpiration, 25
Transportation, 42
  changes, 67
  costs, 44, 57–58, 61, 63, 67, 87
  differences in modes of, 51
  improvement of, 136
  routes, 63
  settlements, 135 (fig.)
Traverse(s), 18–19, 23
Tree(s):
  crops, 141
  rings, 143
  tropical, 40
Trewartha, G. T., 20, 34–35, 38, 73, 1 126 (fig.)
Triangular system, 9
Tribal areas, 81
Tropics, 38, 40, 102, 152, 156
  climatically-determined seasonality in t 39
  decomposition of organic materials in humid, 95
  as the great granaries of the future,
  humid, 154
  inability to settle the, 156
  movement of plants within the, 97
  religions in the, 78
Truck:
  -crop production, 39
  crops, 79
  products, 60
Tschelinzeff, A., 116
Tuan, Y. F., 143
Tubers, 39
Typology(ies), 36, 113, 115, 124, 129
Tysgkiewicz, W., 92n

UNDERDEVELOPED AREAS, 161
  agriculture in, 148
  caloric requirements in, 160
  resources and problems of, 112
Undernourishment, 160–62
Uniform criteria, 21
Unique, the, 3
United Kingdom, 62

nited States, 32, 72, 86, 89, 101, 104, 106, 113, 122–23, 128, 139, 144–45, 152
agricultural geographers in the, 22
Census Bureau, 111
Eastern:
cultivation of the, 65
largest cities in the, 146
farmers of German stock in the, 73
farming rationalization in the, 102
farmsteads, 20
field work, 20
flood plains in the, 149
freight rates in the, 63
fruit regions in the, 115
labor intensity in the, 63
land use patterns in the, 42
"life cycle" of rural population in the, 107
net gain in the, 63
Northeastern, 39, 43, 56, 69
farmers in the, 44
milk production in the, 64
milksheds in the, 119, 136
podzolics in the, 55
readvance of the forests in the, 106
popularity of the land use map in, 23
population ranges for the, 152
regional specialization in the, 64
relationships between Canada and the, 85
Soil Conservation Service, 56
Southeastern:
"bench" terraces in the, 53
land use trends in the, 109
large farms in, 102
types of farming, 20
Western, 22, 88, 135 (fig.), 146
agricultural regions in the, 100
height limits in the, 133
transhumance in the, 42
niversals, 11
psala, 74
rban:
influence, 18
proletariat, 148
sprawl, 146
taxes, 145
Jrban shadow," 144
rbanization, 102
anticipation of, 145
losses of the best soils to, 144
ruguay, 160
.S.S.R. *See* Soviet Union

ACATION, 40
allaux, C., 7
alue(s):
added by farming, 22
"energy," 161
"protective," 161
Value added," 119
an Cleef, E., 75n
an Royen, W., 26n
an Valkenburg, S., 85, 123n, 157
anderhill, B. G., 89, 131–32
Variables, 11
Veatch, J. O., 114
Vegetable(s), 40, 60, 79, 101–2
farms, 37, 128
raising, 38
seed-sown, 39
Vegetation, 121
Vehicles, 133
Verhasselt, Y., 91
Vertical zonation, 35, 131
Veyret, P., 34n, 36
Vidal de la Blache, P., 8, 10
Village(s):
farm, 77, 80–81, 98, 105, 108
industrial establishments in, 148
influence of religion on, formation, 78
social amenities of the, 61
society, 81
Vincennes, 101
Vine, 78, 101–2
Vineyards, 100, 102
"Virgin Land" reserves, 130
Visher, S. S., 7, 25, 31, 32n, 113, 156–57, 159 (fig.)
Vitamin deficiencies, 161
Vitamins, 160
Vogt, W., 141, 147, 156

WAGEMANN, E., 153, 154n
Wagner, P. L., 8, 9n, 78n
Wagret, P., 54n, 89n
Waibel, L., 2, 5–6, 13, 69n, 73–74, 75n, 84–85, 121–22
Waikato, 128
Wales, 65, 118
Ward, J. T., 145
Warkentin, J., 77, 105
Warntz, W., 13
Wars, 143
Wartenberg, C. M., 59
Wasteland, 22
Water, 45, 47, 56, 98, 121
extracting, 142
soil-, capacity, 146
supplies in arid regions, 146
table(s), 25, 142
transferral, 88
Waterways, 99
Watson, J. W., 85n, 91
Weaver, J. C., 21, 22n, 27, 29, 31, 32n, 36, 115, 125
Webb, W. P., 88
Wehrwein, G. S., 59
Weir, T. R., 35
Wells, 109, 130
Werth, E., 100
West, the, 22, 146
West Africa, 84
West Germany. *See* Germany
West Indies, 80, 100
West Pakistan, 88
Western Pacific, 51
Wet lands, 88
Wheat, 21, 32–33, 60
extensive reductions in acreages of, 86
winter-, region, 149
Wheatlands, 128
Wheatley, P., 53
Whitaker, J. R., 141–42
Whitbeck, R. H., 7
White(s), 73, 155
prejudice, 74
settlement, 152
White, C. L., 42, 154n
Whittington, G., 53n
Whittlesey, D. S., 17, 24, 35n, 84, 100, 120, 121n, 123–24, 136
Whittow, J. B., 150n
Whyte, R. O., 95n
Wibberley, G. P., 145
Wilken, G. C., 47, 48n
Wilson, A. W., 146, 147n
Wind erosion, 140. *See also* Erosion
Winterberg, A., 91
Wisconsin, 34, 64, 104, 128, 140
cheese factories in, 106
cheese regions in, 122
foreign cheese industry in, 73
German settlement in southeastern, 74
-Illinois border, 64
service areas in southwestern, 120
Wissmann, H. von, 94
Wolfanger, L. A., 55
Wolpert, J., 29n
Wood, P. D., 150n
Wood, W. F., 19
Woodland, 132
Work cycle, 37
Worker(s):
agricultural, 83, 157
plantation, 40, 83
World Land Use Survey, 26, 157
World War II, 20, 40, 118, 144, 148
Woytinsky, E. S., 25, 56
Woytinsky, W.S., 25, 56

YAMATO BASIN, 35
Yang, M. C., 81
Yield intensification, 43
"Yield relative," 161
Young, A., 1
"Young pioneer area," 133
Yucatan, 105

ZELINSKY, W., 23, 99, 104, 107, 108n, 154n, 157n
Zobler, L., 27–28, 29n, 161
Zonation. *See also* Classification; Regionalization
of agricultural uses around a city, 145
of environmental conditions, 65
of intensity, 67
inverted, 67
market milk-cheese-butter, 65
Thünen-type, 69
vertical, 35, 131
worldwide pattern of, 66
Zoning:
agricultural, 145
ordinances, 87
restrictive, 67, 145
vertical, 131
Zoology, 34